Writing the
Laboratory Notebook

Writing the
Laboratory Notebook

Howard M. Kanare

AMERICAN CHEMICAL SOCIETY
WASHINGTON, D.C.
1985

Library of Congress Cataloging in Publication Data

Kanare, Howard M., 1953–

Writing the laboratory notebook.

Bibliography: p.
Includes index.

1. Laboratory notebooks.
I. Title

Q180.58.K36   1985   542   85–15606
ISBN 0-8412-0906-5
ISBN 0-8412-0933-2 (pbk.)

About the Author

Howard M. Kanare received his B.S. in chemistry from the University of Connecticut in 1976, followed by an M.S. in materials science from the University's Institute of Materials Science. He worked in specialty chemical production before joining the Research and Development Laboratories of the Portland Cement Association, where he now holds the position of Senior Research Chemist. His recent work has included improving X-ray fluorescence analysis of minerals and supervising the production of cement standard reference materials under contracts with the National Bureau of Standards.

"Live in the serene peace of laboratories and libraries."

Louis Pasteur (1822-1895)—Advice to young scientists attending the Jubilee Celebration in Pasteur's honor, at the Sorbonne, 27 December 1892.

I dedicate this book with love to my wife, Larina Veronica Kanare.

Contents

Preface

This book has a simple purpose: to teach the principles of proper scientific notekeeping. Few, if any, of us working scientists write notes as carefully and completely as we should. We often lament not recording a seemingly unimportant detail that later proves crucial. Much experimental work could have been better understood, and much repetition of work avoided, if only we were more attentive in our notekeeping.

The principles presented in this book are goals for which we must strive; sometimes following all the principles will be impractical or impossible. Do the best that you can, and keep in mind that the extra effort and time put into notekeeping are worthwhile.

Everyone develops a personal notekeeping style; some are more efficient than others. If you would like to share the personal techniques that make your notes especially useful, please send your ideas to the ACS Books Department, 1155 16th Street, NW, Washington, DC 20036.

Acknowledgments

The comments, criticisms, technical information, and suggestions of many people have been invaluable. I would like to thank the following people for their assistance in the development and revision of this book: Donald W. Anstedt, Construction Technology Laboratories; Paul Banks, Columbia University; Henry A. Bent, North Carolina State University; Walter E. Buting, Genentech, Inc.; George J. Collins, U.S. Government Printing Office; Raymond E. Dessy, Virginia Polytechnic Institute and State University; Hubert E. Dubb, Fleisler, Dubb, Myer, and Lovejoy; Edmund L. d'Ouville, formerly with Standard Oil Company (Indiana); Merrily Smith, Library of Congress Preservation Office; John K. Taylor, National Bureau of Standards; Richard C. Toole, DuPont Experimental Station; and William K. Wilson, formerly with the National Bureau of Standards.

I acknowledge a personal debt of gratitude to two of my professors, John T. Stock and Bertrand L. Chamberland, who, years ago, taught me the virtues of proper notekeeping. Since that time, many of my colleagues in industrial research positions have confirmed the value of careful notekeeping, especially in regulated environments. Their comments and suggestions are appreciated.

Howard M. Kanare
Chicago, Illinois

May 20, 1985

The Reasons for Notekeeping

An Overview

A laboratory notebook is one of a scientist's most valuable tools. It contains the permanent written record of the researcher's mental and physical activities from experiment and observation, to the ultimate understanding of physical phenomena. The act of writing in the notebook causes the scientist to stop and think about what is being done in the laboratory. It is in this way an essential part of "doing good science."

The laboratory notebook is known by different names to different people. It is variously called the "lab notebook," "project notebook," "engineer's log book," or "research journal." Whatever it is called in your place of work or study, I am referring to a bound collection of serially numbered pages used to record the progress of scientific investigations. The notebook is used not only in the laboratory, but also at pilot plants and in field studies.

Why Keep a Notebook?

The information written into a research notebook is used for several purposes. Most importantly, the pages of the notebook are used to preserve the experimental data and observations that are part of any scientific investigation. The notes must be clear, concise, and complete. The properly kept notebook contains unambiguous statements of "the truth" as observed by the scientist. If the notebook is to be of any value, experiments that fail must be recorded as faithfully as those that succeed.

The guiding principle for notekeeping is to write with enough detail and clarity that another scientist could pick up the notebook at some time in the future, repeat the work based on the written descriptions, and make the same observations that were originally recorded. If this guideline is followed, even the original author will be able to understand the notes when looking back on them after considerable time has passed!

0906–5/85/0001/$06.00/1
©1985 American Chemical Society

After the data are recorded, the researcher begins to study them. The notebook provides a forum in which data and observations are analyzed, discussed, evaluated, and interpreted. Even though much experimental data is today in printed instrumental output or in computer-readable form, the notebook is still the logical place where such data are summarized and reviewed. This process leads to the writing of reports, technical papers, patent disclosures, and correspondence with colleagues. If the notebook is well organized and adequately indexed, finding the appropriate passages when needed is easy without searching page by page through the notebook.

The information recorded in the notebook also can be used to review progress and to plan further work.

The Industrial Notebook

From an industrial research scientist's standpoint, keeping a notebook is an integral part of the job. The lifeblood of many corporations is the steady stream of patents and inventions flowing from research and development laboratories into the marketing department. The important role that a notebook can play in securing or protecting a patent is discussed in Chapter 7. If properly kept, the notebook is the scientist's proof of discovery or invention and often is the sole source of certain information that will be needed in pursuit of a patent. Notebooks should not be kept according to certain standards solely to ensure proof of discovery or invention, or to get a patent. Rather, if the notebook is kept according to good principles, it will be a useful device in any research, and especially so when patents are involved.

From the research supervisor's standpoint, occasional review of the employee's writing can help to assess progress. This review should feed information back into the process of improving the researcher's effectiveness. This aspect applies in almost all job situations and is unfortunately neglected by many supervisors. Checking of primary data and calculations by a colleague or supervisor is often required by research protocols and is recognized as an essential step to guard against fraud in research (*see* Chapter 4).

The Academic Notebook

The student who is beginning laboratory course work will have to keep a notebook because it is required for the course. The student should realize that good notekeeping is an acquired skill that can be of tremendous benefit in any career. If this skill is well developed while in school, notekeeping will become a matter of habit rather than a chore. *Writing good notes requires discipline and practice.* It is a skill that does not come easily to most people.

For the student performing original research, the notebook will be the prime source of information required to write a thesis, dissertation, or technical paper for publication. Careful notekeeping while performing the dissertation research can make writing of the dissertation much easier and may help to avoid repeating certain experiments. The advanced student may even become an inventor, and the notebook may play a pivotal role in obtaining legal right to the invention.

Why Use a Bound Notebook?

The foremost reason for using a bound notebook rather than a loose-leaf binder or wire-spiral notebook is that the pages are permanently and strongly attached together. Related notes written on different pages will not be separated by accidental loss of separate pieces of paper. In addition, the date of a particular entry is less subject to question if notes are recorded in a consistent style, in chronological order, and with no blank or missing pages. The industrial researcher, whose work may lead to patents, has no choice except to use a bound notebook for all laboratory notemaking.

Tales of Fame and Fortune

The number of discoveries or inventions that have been delayed or missed because of sloppy notekeeping can never be known. The following story demonstrates how a great discovery passed by the French astronomer, Pierre LeMonnier (1715–1799), because he kept poor notes.

Sir William Herschel (1738–1822) discovered what he thought was a comet in March 1781. The heavenly body he observed was soon shown to be a new planet and was given the name Uranus. However, Uranus was observed on several occasions back in the 1760s by LeMonnier, who didn't realize that he was looking at an undiscovered planet. Francois Arago (1786–1853), a French astronomer and biographer of the nineteenth century, was studying the moon's librations from the Paris Observatory in 1805. During this time, his colleague, Alexis Bouvard (1767–1843), showed him LeMonnier's notes. Arago wrote (1)

> If the writings were held in order, if the determinations of right ascension and declination for each day were figured in regular columns and observed, a simple glance would have shown to LeMonnier, in January 1769, that he had observed a moving heavenly body, and the name of this astronomer, instead of the name Herschel, would be found forever next to the name of one of the principal planets of the solar system. But LeMonnier's records were the picture of chaos. Bouvard showed me, at the time, that one of the observations of the planet Uranus was written on a paper bag which once upon a time held powder to powder the hair, bought at a perfumer's....LeMonnier, in reviewing his records, found that he had observed three times the so-called comet of Herschel in 1763 and 1769.

Most lab workers, from time to time, may use the odd bit of paper at hand to jot down a tare weight, titrant volume, or instrument reading. Often an idea will occur to you and you want to put it on paper before the thought is lost, but your notebook is not around. Henry Ford always carried a small note pad in his suit coat pocket. When he came across a phrase he especially liked or an idea for a new device or a method of improving a mechanical process, he would immediately write down his thoughts. Dozens of pieces of paper from these little "jot books" were found in small boxes in his home, Fair Lane, shortly after his death in 1947. Many words were terribly misspelled and none of the notes was signed or dated. These notes provide some interesting insights into the thoughts of the great industrialist, but they would have been worthless in proving the date of any particular invention. (Acceptable methods of transferring notes into the permanent notebook are discussed in Chapter 5.)

Perhaps the most famous and potentially valuable notebooks of recent decades were those written by Gordon Gould concerning the invention of the laser (2). Gould was a graduate student at Columbia University in 1957 working just down the hall from Charles H. Townes, who was then a professor in the physics department. In April of 1951, Townes realized the conditions that would be necessary to get stimulated emission of microwave radiation, a concept that would be turned into a working *maser* within three years. In September of 1957, Townes sketched his idea for an optical maser. He then discussed his idea with his brother-in-law, physicist Arthur Schawlow, who had also taught at Columbia but was now at Bell Labs. Townes and Schawlow developed a detailed plan for building the device that would eventually be known as the *laser*.

Gould was inspired by Townes' invention of the maser. In November of 1957, Gould sat down with his notebook and wrote down his own idea about how to build an optical maser. He coined the word *laser*, which stands for light amplification by stimulated emission of radiation. He recognized that a laser would be a nonthermal source of light that could heat objects to extremely high temperatures and could be used, for example, to melt steel. Gould also proposed that laser light could be used to initiate thermonuclear fusion and to communicate between the earth and the moon. He took his notes to a Bronx candy store owner, who notarized them.

About that time, Townes spoke with Gould about Gould's thesis research, which was on a different but related subject. Townes said that he also explained what he was doing to Gould, and Gould admits that the conversation warned him of what Townes was planning. The two men disagree, however, over when the discussion took place. Gould states that it happened after he had written his notes on the laser, and Townes claims that it was before the notes were written.

Many other physicists as well as the government of the United States were involved in the subsequent drama surrounding the laser. Townes and Schawlow eventually received patents on the laser and Nobel prizes for their work in physics. Gould began many years of expensive and frustrating legal proceedings to invalidate some of these patents and to get patents for himself. He was finally issued a patent in 1977 on optical pumping, a technique used in many lasers. Gould has reportedly realized over two million dollars in proceeds from the sale of his interest in the invention, but the victory is bittersweet. His patent on optical pumping is being contested and the outcome is uncertain.

What steps could Gordon Gould have taken that might have helped to establish his priority as inventor of the laser? First, he should have had his notebooks witnessed by a competent physicist, who could testify as to the date and conditions under which Gould claimed to have made the invention. Second, Gould could have made a public disclosure of his invention by publishing his ideas in a technical journal as a defensive measure. Third, he could have recorded in his notebooks the details of his conversations with Townes immediately after they took place. None of these steps could ensure victory in such matters, but a properly witnessed notebook is an important step in establishing credibility.

For examples of outstanding notebooks, look to outstanding scientists. Linus Pauling's notes (3) are the epitome of grammatical, clearly organized thoughts and calculations. According to W. I. B. Beveridge (4), "The notes kept by Louis Pasteur afford a beautiful example of the careful recording of every detail." Michael Faraday, the self-taught genius who is the only person to have two units of measurement named after him (the farad and the faraday), wrote careful notes of every experimental step (5). The pages of his notebooks contain many detailed drawings of his apparatus.

Meeting Some Legal Obligations

The practice of keeping laboratory and research records in a bound notebook is more a matter of tradition and good sense than a matter of obeying any laws. Major government agencies that provide the bulk of research funding at universities (including the National Science Foundation, National Institutes of Health, and the U. S. Department of Energy) have issued detailed regulations for cost accounting and reporting progress in research. Surprisingly, however, they have virtually no requirements for the documentation of experimental work or logging of original data. Although no such requirements exist, everyone, from student to corporate research leader, can benefit from a well-organized system of notekeeping and notebook management.

Although no government laws specifically require researchers or analysts to keep handwritten records in bound notebooks, the proper use of a notebook is often the best means of meeting some legal requirements. Perhaps the most heavily regulated field of scientific research is the pharmaceutical field. Because pharmaceutical products affect the lives of humans, pharmaceutical researchers must prove the safety and efficacy of their products. To this end, regulations exist for the recording and checking of data generated in the pharmaceutical laboratory. The U.S. Food and Drug Administration has issued "Good Laboratory Practice for Nonclinical Laboratory Studies" (21 CFR, part 58), which requires that "All data generated during the conduct of a nonclinical laboratory study, except those that are generated as direct computer input, shall be recorded directly, promptly, and legibly in ink. All data entries shall be dated on the day of entry and signed or initialed by the person entering the data."

Strictly speaking, these guidelines apply only to a *nonclinical laboratory study* that assesses the safety of a drug, food additive, or medical device for human use. The guidelines do not apply to any test using human subjects or to any exploratory studies carried out to determine the physical or chemical characteristics of a test article. However, the standard practice in pharmaceutical research facilities is to use bound notebooks for recording the progress of nearly all types of investigations. In addition to the notebooks, many documents that are created during the course of these studies are kept in project files. These laboratories rely heavily on computers to generate data and to keep track of work (the audit trail). Some pharmaceutical lab managers have stated their intent to have a paperless, electronic audit trail by 1990.

In a very different field, the American Society for Testing and Materials (ASTM) is a voluntary organization whose members come from government, industry, and academia. They have diverse backgrounds in manufacturing, construction, and materials testing. ASTM publishes voluntary consensus standards (6) that can be adopted by any organization as part of a laboratory protocol or as part of a business contract. In 1982, ASTM issued a "Standard Guide for Accountability and Quality Control in the Chemical Analysis Laboratory" (ASTM E882-82), which suggests methods of identifying samples and maintaining records. The intent of parts of this standard could be met by using bound notebooks, although the standard does not specifically require the use of bound notebooks for recording analytical data.

Another example of the use of bound notebooks is provided by the American National Standards Institute document designated ANSI/ASME NQA-1-1983 Edition, "Quality Assurance Program Requirements for Nuclear Facilities." A section of this document states that "Records that furnish documentary evidence of quality shall be specified, prepared, and maintained. Records shall be legible, identifiable, and retrievable....Documents shall be considered valid records only if stamped, initialed, or signed and dated by authorized personnel or otherwise authenticated."

Properly kept notebooks will meet some of the requirements of this standard even though the standard does not specifically call for the use of handwritten, bound notebooks. Many quality assurance (QA) documents are based on printed forms, which are filled out and stored in file folders or binders. However, the original data generated in laboratory work, such as chemical analysis or physical testing of materials, is best preserved in the individual tester's notebook. QA guidelines based on this and other standards are being adopted by an increasing number of organizations, including many outside the nuclear power industry.

Other Types of Laboratory Records

A scientist's laboratory notebook is part of an information system that embraces instrument and sample logbooks, project files, purchase records, and many other items. All of these elements should contribute to the efficiency of a scientist's work, although, at times, just the opposite may seem to be true! For the purpose of this book, we adopt the following distinction between notebooks, logbooks, and diaries:

- A *notebook* is a bound collection of blank, serially numbered pages used to record the progress of scientific investigations. Projects recorded in notebooks include basic and applied research, product development and evaluation, "thought experiments" and calculations, and the conception and reduction to practice of inventions. A notebook is usually assigned to a single person and may have recorded in it more than one project. Entries occasionally may be made by other people.

- A *logbook* is a strictly chronological listing of events such as the handling of samples, instrument usage, equipment performance, or labor records. Each logbook is usually used for just one of those purposes. Such a logbook usually will have daily entries, which might be handwritten into a bound book or onto printed forms. These forms are then collected in a binder or folder. A logbook is often kept in a fixed location (e.g., near an instrument) and may have entries made by many people. Logbooks for samples and instruments are described in "Standard Guide for Records Management in Mass Spectrometry Laboratories Performing Analysis in Support of Nonclinical Laboratory Studies," ASTM E899–82.

- A *diary* or *personal journal* is maintained by an individual. It comprises a synopsis of each day's work and is intended to give a simplified, overall indication of how the day was spent. The diary can have personal insights, opinions, and comments on any subject, compared to a notebook, which must be a factual account of work.

Computer-based laboratory information management systems (LIMS) have become commonplace. These systems supplement or replace handwritten records in a variety of applications. The status and future of computer notekeeping is discussed in Chapter 8.

A Proper Notebook Page

The purpose of this book is to teach the skills needed to write the most useful notes. In subsequent chapters, we will explore how to make proper notekeeping a habit and how to get the most value for the time that you spend writing your notes. So as not to keep you in suspense any longer than necessary, the essential features of a proper notebook page are illustrated by the example in Figure 1.1 and summarized in the following list.

- The entry was written immediately after the work was performed.
- The author has dated and signed the entry.
- Each section has a clear, descriptive heading.
- The writing is legible and grammatical.
- The use of the active voice in the first person tells the story and clearly indicates who did the work.
- The entry was read by a witness, who signed and dated the page.

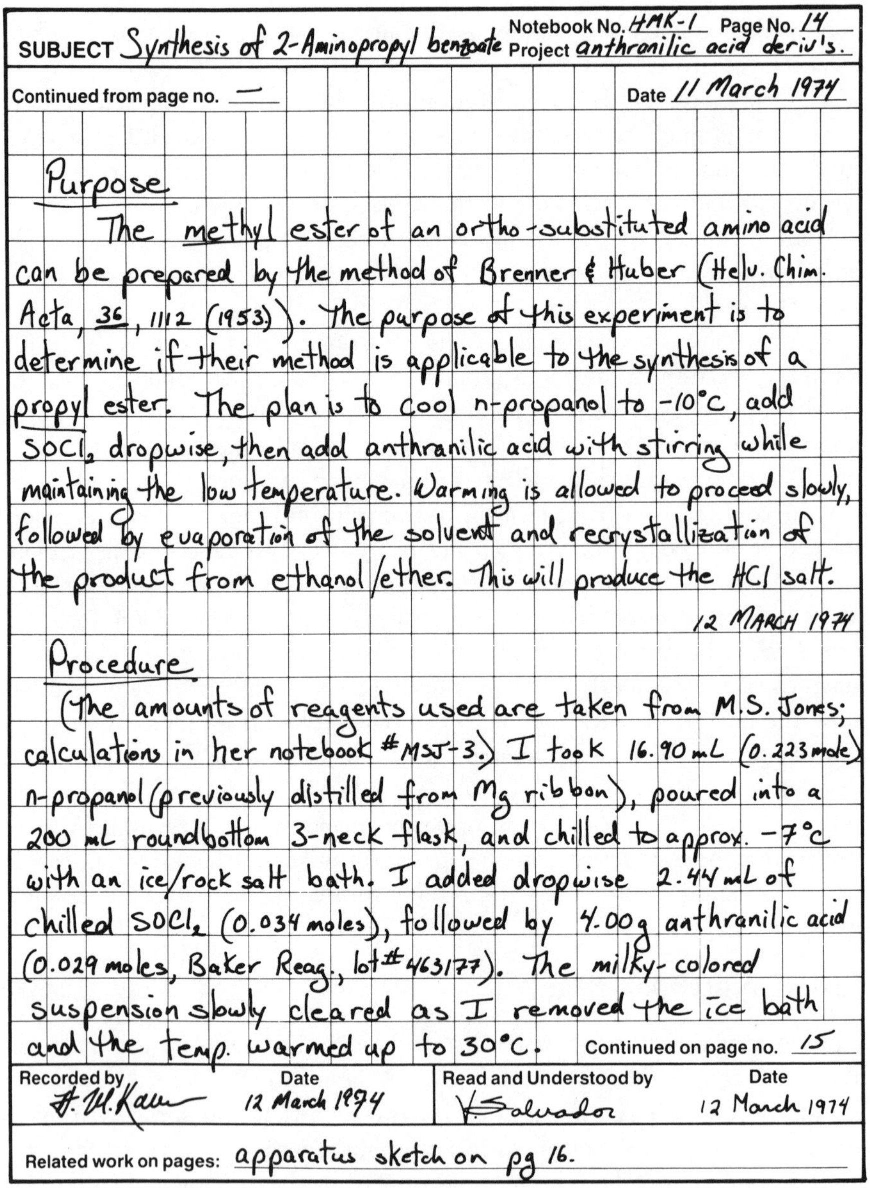

SUBJECT _Synthesis of 2-Aminopropyl benzoate_	Notebook No. _HMK-1_ Page No. _14_ Project _anthranilic acid deriv's._

Continued from page no. —   Date _11 March 1974_

Purpose

The <u>me</u>thyl ester of an ortho-substituted amino acid can be prepared by the method of Brenner & Huber (Helv. Chim. Acta, <u>36</u>, 1112 (1953)). The purpose of this experiment is to determine if their method is applicable to the synthesis of a propyl ester. The plan is to cool n-propanol to -10°C, add $SOCl_2$ dropwise, then add anthranilic acid with stirring while maintaining the low temperature. Warming is allowed to proceed slowly, followed by evaporation of the solvent and recrystallization of the product from ethanol/ether. This will produce the HCl salt.

12 MARCH 1974

Procedure

(The amounts of reagents used are taken from M.S. Jones; calculations in her notebook #MSJ-3.) I took 16.90 mL (0.223 mole) n-propanol (previously distilled from Mg ribbon), poured into a 200 mL roundbottom 3-neck flask, and chilled to approx. -7°C with an ice/rock salt bath. I added dropwise 2.44 mL of chilled $SOCl_2$ (0.034 moles), followed by 4.00 g anthranilic acid (0.029 moles, Baker Reag., lot #463177). The milky-colored suspension slowly cleared as I removed the ice bath and the temp. warmed up to 30°C.

Continued on page no. 15

Recorded by _F. M. Kaw_	Date 12 March 1974	Read and Understood by _V. Salvador_	Date 12 March 1974

Related work on pages: apparatus sketch on pg 16.

Figure 1.1 A page from a properly kept notebook.

Notes and Literature Cited

1. Arago, Francois *Astronomie Populaire*; Paris: 1857; p. 489. (Translated from French by H. M. Kanare.)
2. Hecht, J. and Teresi, D. *Laser: Supertool of the Eighties*; Ticknor and Fields: New York, 1982. (*See* this reference for a more in-depth version of the laser story.)
3. Linus Pauling's notebooks relevant to the development of quantum theory are on microfilm at the American Philosophical Society, Philadelphia, PA.
4. Beveridge, W. I. B. *The Art of Scientific Investigation*; W. W. Norton: New York, 1951; p. 17.
5. Faraday's research notes were written in bound notebooks and on separate sheets of paper, which he gathered in folios. The notes were typeset, along with his drawings, and were published as *Faraday's Diary*; Martin, Thomas, Ed.; G. Bell and Sons: London, 1932. The eight volumes comprise more than 4000 pages and cover the years 1820–1862.
6. ASTM publications are available from the American Society for Testing and Materials, 1916 Race Street, Philadelphia, PA 19103.

The Hardware of Notekeeping

Books, Pens, and Paper

This chapter is intended especially for the laboratory manager, purchasing agent, or school administrator who is responsible for specifying and obtaining laboratory notebooks, writing paper, and pens. The information that is included will also be valuable to librarians, archivists, record managers, and any scientist who is concerned about the permanence of written records.

Longevity of Research Notes

From a legal standpoint, original notes of inventions or discoveries should be available for use as evidence in patent disputes for 23 years after the date the patent is issued. The U.S. Food and Drug Administration requires that original records of research from nonclinical laboratory studies be maintained for 10 years after completion of the study. As technology advances, old notes become less and less useful from a practical standpoint. In both industrial and academic settings, it is not uncommon to look at notebooks that are 10–15 years old in an attempt to determine if a particular course of research was ever tried. Occasionally, notebooks 30 years old or more furnish some practical information; many are certainly of historical interest. Therefore, we should be concerned about maintaining original research notes for at least 25–30 years; the paper should be in such good condition that it can be handled and studied without fear of damage to the physical record. At the same time, the writing must be in such good condition that it can be read and understood without ambiguity. If the paper has darkened or the ink has faded, information may be compromised.

Permanent Paper

Sadly, many scientific notebooks written between the 1930s and 1960s have deteriorated badly. This problem stems from two causes: the use of light-sensitive

0906–5/85/0011/$06.00/1
©1985 American Chemical Society

inks, which fade rapidly when exposed to sunlight; and the use of cheap, unstable paper, which degrades even when it is stored in drawers or cabinets and protected from sunlight. We now have the knowledge and technology to manufacture and use paper and ink that will truly last for several times the lifetime of the author.

Why are many paper documents created hundreds of years ago in better condition than those made as recently as 50 years ago? The answer lies in two aspects of papermaking technology that occurred in the second half of the nineteenth century (1). First, alum–rosin compounds began to be used as paper sizing agents after the Civil War. Sizing agents are materials that are used to decrease the natural tendency of paper to soak up moisture or ink. Starch and synthetic polymers are commonly used for this purpose today. Unfortunately, alum–rosin compounds react with atmospheric moisture to produce acids that attack the main components of ordinary paper, the cellulose molecules, and, thus, weaken and embrittle the paper. Second, as an economic measure, mechanically ground wood began to displace chemically purified wood pulp and other pure cellulose sources (e.g., cotton rag fibers) in the 1870s. The short fibers in ground wood produce a weaker paper than the long cellulose fibers from chemically treated wood pulp. In addition, the lignins in ground wood degrade to form acids that attack the cellulose molecules. Ground wood is also more susceptible to decomposition in sunlight than chemically purified wood pulp. Modern newsprint is an example of paper made with a large percentage of ground wood.

Permanent paper is made from 100% chemically purified wood pulp, contains no ground wood or lignin, and does not have the alum–rosin sizing agent. Such paper will have a pH value close to neutral. Paper with the greatest stability is also buffered with several percent of calcium carbonate as an alkaline reserve that is available to react with and neutralize acids produced during degradation of trace components in the paper. This alkaline reserve also protects the paper from attack by atmospheric acids caused by pollution.

Not much has been said so far about rag content, which everyone commonly associates with high-quality paper. Actually, increased rag content (i.e., cellulose fiber derived from cotton rags, linen, or unused industrial waste) produces improvement in the physical properties of paper such as folding endurance and tear resistance. Long, strong, purified wood fibers have the same effect. A high content of rag or other long fibers does not guarantee permanence if the paper contains impurities that will degrade the cellulose molecules. In fact, decreased acidity (pH 5.5 or greater) has been well correlated with increased permanence. If paper is chemically pure for permanence and contains sufficient rag content or chemically purified wood pulp to be physically durable, it is referred to as *permanent-durable* paper.

The highest quality, permanent-durable writing paper produced in 1984 and stored properly is expected to last 500 years. Some paper experts are even more optimistic and believe that under proper conditions such papers will last indefinitely. Although buying paper that will outlast the usefulness of the notes by many generations may seem to be a needless expense, such paper will certainly remain in nearly pristine condition for as many years as the notes are needed. The researcher who writes notes with archival quality paper and ink can be comfortable knowing that the paper will be strong and the ink clearly legible for as many years as the notes might be useful.

How can you tell if the paper in your notebooks is of this permanent type? The notebook supplier may be willing to check production records and determine the source and type of paper used in the books. If not, you can do a few simple spot tests on sample pages of paper using an inexpensive paper testing kit. (See the supplier list at the end of this chapter.) The spot tests will determine the presence or absence of alum–rosin compounds, lignin, and starch, and can be used to estimate the pH of the paper. A competent microscopist must be called to determine the type of fibers present and may be needed to arbitrate the presence or absence of high-yield wood pulp or ground wood.

In addition to these general guidelines, you may use ASTM Standard Specification D 3290, "Bond and Ledger Papers for Permanent Records" (2). This standard provides for three levels of permanence called Types I, II, and III; each type is divided into two categories of use, ordinary and high referral. Type III paper has a minimum pH of 5.5, is free of ground wood, and is expected to last at least 50 years. Type II paper has a pH value between 6.5 and 8.5 and is expected to last in excess of 100 years. Type I paper has an alkaline reserve of calcium or magnesium carbonate and a pH value between 7.5 and 9.5. This type of paper is expected to last several hundred years.

Chemical research laboratories in which notebooks are used often have mildly corrosive atmospheres because of occasional spills and the vapors of reagents, despite the lab designer's best intentions. Undoubtedly, buffered papers will enjoy longer life under such conditions.

Acid-free buffered papers (ASTM Type I) are the most expensive; this type of paper costs approximately two to three times the cost of ordinary paper. Paper with a minimum pH of 5.5 may be purchased for only 50% more than ordinary paper. You get what you pay for.

These costs can be put into perspective if you consider that a typical industrial research project costing $150,000 might involve several people who produce three or four notebooks filled with experimental data. These notebooks will cost around $40 if made with ordinary paper, and approximately $100 if made with acid-free, buffered permanent-durable paper. The difference in cost is 0.04% of the total cost of the project. For this miniscule difference in price, how can anyone not afford to specify and use the highest quality writing paper?

Adverse storage conditions can dramatically shorten the life of even high-quality paper. Control of lighting, temperature, and humidity as well as the cleanliness of storage facilities are all-important factors and are discussed in Chapter 4.

Pencils, Pens, and Inks for Handwritten Notes

For several good reasons, notes should not be recorded with a graphite pencil. Most important, pencil writing can be erased easily and is, therefore, not permanent. Because pencil writing can be erased and the data rewritten, the researcher will be open to questions about the authenticity of data. Writing with permanent ink ensures that another person cannot alter your notes accidentally or intentionally. A second reason to avoid the use of pencil is that constant handling of the notebook causes the pencil writing to become smudged and illegible. Finally, pencil does not photograph well and is therefore unsuitable for microfilm reproduction of the notebook.

Many types of writing instruments are available, including ball-point pens, fountain pens, porous "felt" tip pens, plastic ball roller pens, and mechanical drawing pens. Do these different kinds of pens differ significantly in their performance? Which type is best? Are colored inks as suitable as black ink?

Ink to be used for handwritten research notes should be fast drying, highly resistant to aqueous and organic solvents, light fast (stable toward fading in sunlight), reproducible for microfilming and photocopying, stable during long-term storage, and unreactive toward paper.

Traditional inks that contained gallic acid, tannic acid, hydrochloric acid, and ferrous sulfate were in common use for many centuries before researchers at the U.S. National Bureau of Standards reported in 1937 that such inks attacked and weakened the paper on which they were written (3). As recently as 50 years ago, a U.S. Government specification stated that writing ink should contain hydrochloric acid (4). Such acidic iron-gall and gallotannin inks have actually eaten right through papers to the extent that today some old documents look like lace!

Fortunately, modern inks will not cause such problems. In fact, the inks found in a few brands of ordinary ball-point pens meet some of the criteria for permanence. However, many common inks do not. Essentially, modern ball-point pen inks consist of concentrated dyes in organic solvents such as propylene glycol and benzyl alcohol. In contrast, inks for porous-tip, roller tip, and fountain pens can contain more than 60% water. Aqueous-based inks are generally considered inferior in permanence to solvent-based inks.

Another reason to avoid using porous-tip pens is that the low viscosity ink tends to spread quickly through the paper fibers during writing; thus, the image spreads or bleeds, often through to the other side of the paper. This phenomenon is an interaction of the ink, paper fibers, and chemical sizing agents, which vary from paper to paper. One brand of pen may work well on one type of paper, but bleed badly on another type of paper.

I recently conducted tests of nine popular brands of ball-point pens and six porous-tip pens. These tests indicated that the inks could withstand 800 h of fluorescent light at the intensity normally found in a laboratory or office (70 footcandles) without visible fading. The same inks were then exposed to sunlight, and nine of them faded badly after just 6 h. The red and blue porous-tip pen inks faded the most, followed by some of the blue ball-point pen inks. Black inks from both types of pens were the least affected by sunlight.

Why does color make a difference? Most dyes decompose when exposed to UV and short-wavelength visible light. Sunlight affects red inks the most, black inks the least, and blue inks somewhere in between. The effect of sunlight is less for black and blue inks because most of these ink colors contain phthalocyanine dyes, which are not light sensitive. In addition, some of these inks contain carbon black, one of the most stable pigments known. Although some red inks do contain pigment particles, they are still more sensitive to sunlight than black and blue inks.

Tests of 15 pens revealed large differences in their resistance to various solvents. None of the ball-point pen inks were affected by water or hexane, some bled slightly when treated with dilute hydrochloric acid or with acetone, and most bled badly when treated with methyl alcohol. The behavior of the porous-tip pen inks was similar to the ball-point pen inks.

How should you select a pen for writing permanent laboratory notes and records? Probably the best course of action is to inquire of the pen manufacturers whether or not their pens meet the performance standards set forth in Federal Specification GG–B–60D section 3.5 (5), or in the German Institute for Standards DIN 16 554 part 2 (6). These two standards describe fading, bleaching, erasing, and written performance tests for ball-point pens.

Unfortunately, writing instrument manufacturers do not make a point of advertising the relative permanence and solvent resistance of their pens. Perhaps an informed buying public will be able to nudge them in the direction of certifying the performance of their products, especially when archival permanence is of concern to the user.

One ink manufacturer has produced an archival quality ink for use in ball-point pens but has had little success in selling it, probably because archival permanence of inks is of little concern to the general public. This ink is commercially available in ball-point pens, which can be purchased in large or small quantities from the supplier listed at the end of this chapter.

A ball-point pen with black ink is best for all permanent scientific notekeeping. Some research companies that microfilm their employees' notebooks require that black ball-point pens be used. To minimize fading from sunlight exposure, the notebook should be kept closed when not in use. In addition, the notebook should be placed a safe distance away from work areas where solvents (especially alcohols) are used because one spill can render a page of data illegible.

The choice of fine or medium tip width is a matter of personal preference. Many writers find that fine point pens tend to produce more uniform lines with fewer blobs than medium point pens. With some combinations of paper and ink, gradual spreading or migration of ink within the capillary structure of the paper is observed. This effect may be minimized by using a fine point pen.

A note on the shelf life of pens: Carbon black and other pigments sometimes coagulate or settle out of suspension during long-term storage even though the inks in ball-point pens are in the form of pastes. This segregation of solvent and pigment can lead to skipping, blobs, and other performance problems. Therefore, it is a good idea to stock only the quantity of pens that will be used within a few months, rather than to buy a supply to last for several years.

Bound Notebooks

In case binding, the most common method of manufacturing the hard-cover notebook, the pages are sewn and glued together; this method is in contrast to mechanical binding, which holds the pages together with a springy plastic binding or a thin wire. The latter type of book is familiar to students as the ubiquitous spiral notebook. Case binding produces a book whose pages are strongly attached and protected by a rigid, durable cover.

During manufacture, the pages for a case-bound book are gathered in folded sets called signatures. The pages are then stitched through their folds and the signatures are sewn together. This process is called Smyth sewing, and it allows the book to be fully opened so that the pages can be laid flat.

Usually, the cover is stiff cardboard, called pasted board, covered with a fabric or a thin, chemically treated paper. Woven cloth and leather coverings are found infrequently; these covers have been replaced with synthetic, nonwoven polymeric materials. The most popular style of cover for industrial research notebooks is pasted board covered with vinyl-like material and cut flush with the pages. Custom notebooks produced for industrial research facilities usually have their covers hot-stamped or imprinted with a unique serial number and the name of the research facility. Lower cost notebooks produced in large quantities for students typically have just a plain paper label on the front cover; this label can be filled in with the student's name and course number.

Notebooks bound with wire staples through the paper folds (called saddle stitching) and notebooks with soft paper covers are also available; these types are not recommended because they are less durable. Plastic-bound books, wire-spiral books, and loose-leaf binders are not acceptable for writing permanent laboratory notes because pages can be intentionally inserted, removed, or accidentally ripped out.

Selection of the size, shape, and number of pages in the notebook should be based on the work at hand. Pocket-size notebooks containing 50 sheets measuring 5×8 in. (127×203 mm) are especially useful for recording data when a few observations are made repetitively over a long time. For example, each page can be set up in a standard format and the data entered, one page for each session. Then, successive observations can be easily taken from the notebook for the interpretation phase of the study. A standard format for repetitive work minimizes the chance for transcription errors.

For research projects lasting several months or years, a notebook containing 100 to 150 sheets measuring $8\frac{1}{2} \times 11$ in. (216×280 mm) is probably the most convenient. This size provides plenty of room on each page for notes, calculations, drawings, and discussion of results. It is not unusual for a scientist to fill several notebooks of this size in a year.

Sometimes samples are inserted into the notebook such as swatches of fabric, thin metal coupons, and other flat samples. An accumulation of such objects can soon lead to distortion and breaking of the notebook's binding. Custom-made notebooks can be produced that can accommodate such samples. During the binding process, cardboard sheets are inserted between some pages and kept in place until the binding is complete. The cardboard inserts are then slipped out, and extra room is left in the binding for the thickness of the expected samples. Your notebook supplier should be asked for this special feature if you expect to put samples into the notebook.

Page Formats

Stock notebooks are available with many styles of grids, lines, and type matter printed at the top and bottom of each page. Custom notebook pages can be produced with any imaginable format that will help the researcher or analyst work more effectively. Supervisors and purchasing agents should periodically seek the advice of the scientists who will use the notebooks to be sure that the product selected is the most suitable for the job to be done. The following features should be kept in mind when choosing stock notebooks or when designing custom notebooks.

The first page in the notebook should contain a short form for recording the date that the notebook was issued, to whom it was issued, and for what purpose (Figure 2.1). This page should also contain space to note when the book was completed and returned for storage. Space should be available to note if this book is a continuation of work recorded in a previous notebook.

The next page or two should contain printed instructions outlining the organization's rules for notekeeping (*see* Figures 4.1-4.3). Although the same principles for good notekeeping apply to all companies, each organization has its preferences for emphasizing particular aspects of notekeeping.

Following the page of instructions, the book should have several pages for a table of contents (Figure 2.2). This format is set up to include the date of entries, subjects, and page numbers. How many lines should be allocated for the table of contents? A rule of thumb is to allow one entry for each page in the body of the notebook. If the table of contents is any longer, the user might just as well flip through the pages to find the entry itself. The table of contents should be a concise list of the dates and projects recorded in the book. If this list is kept up-to-date, it will be invaluable for rapidly finding the location of specific entries a few weeks or even years after they are written.

(Print clearly—use black ball point pen.)

This notebook no. _____

issued to _____
　　　　　　(print name)

_____
(signature)

_____
(title)

_____
(department)

on _____
　　(day)　　　(month)　　　(year)

to be kept in _____
　　　　　　　(location)

Issued by _____
　　　　　　(print name)

_____
(signature)

(Upon completion, fill in the following)

No further entries after _____
　　　　　　　　　　　(day)　　　(month)　　　(year)

Received for storage on _____
　　　　　　　　　　　(day)　　　(month)　　　(year)

Received by _____ _____
　　　　　(print name)　　　　　　　(signature)

Related work continued in notebook no. _____

Figure 2.1 An example of an issuance page, which should be the first page in every notebook. It can be handwritten if not typeset. Especially important are the spaces in which to record any other notebooks that contain records of related work.

TABLE OF CONTENTS		Notebook No. _____	
Date	Subject		Page No.

Figure 2.2 A typical blank table of contents page. See Figures 5.1–5.3 for methods of modifying the table of contents to meet specific needs.

After the table of contents are the pages used for actual notekeeping (Figure 2.3). Each page should be numbered at the upper outside corner. Printed headings at the top of each page remind the writer to fill in essential information such as the date, subject, or project number, and from what page the work is continued. The bottom of the page contains spaces for author and witness signatures and a place to note on what page the work is continued. The writing area can be ruled in many ways, the most popular being a square grid with fine lines spaced $\frac{1}{4}$ in. (6.3 mm) apart. This particular cross-ruling makes it easy to put data in neat columns and rows and to make scale drawings and graphs. Other methods of ruling the notebook pages are shown in Figure 2.4. Notebooks can be ordered with combinations of these page formats. For example, one layout may consist of simple horizontal ruling for writing text on right-hand pages and a grid for aid in plotting data on left-hand pages.

Side margins are typically ruled 10 mm from the edges of the page and writing must not continue beyond these margins if the notebook pages are to be microfilmed for preservation. The lack of wide margins will also help to discourage the writer from developing the bad habit of making marginal notes.

In a stock or custom-designed notebook, sufficient space must be available after each typeset heading. Let us look at the typeset material in more detail, beginning at the top of the notebook page (Figure 2.3).

The date space should have room for the day, date, month, and year: Thu 2 Aug 1984. (When written this way, the date is unambiguous and quickly grasped.)

The subject description needs room for only four or five words. This description should be precise and pertinent to the particular phase of the work recorded on that page because this entry will often be scanned when searching for old data. Space should be adequate for the entire project code number to be written easily without unduly squeezing each character. The writer should try "filling in the blanks" several times on dummy pages to see if headings are laid out well.

At the bottom of the page are places for the author's signature, the date, and the signatures of witnesses. Two phrases are commonly used to precede the witnesses' signatures depending upon the conditions of witnessing. The term "This page read and understood by me" indicates just what it says. It is a good idea to have this phrase on each page although a witness may not actually read and sign each entry unless the work is deemed of suitable importance. An alternative wording might be "Pages _____ through _____ read and understood by me," followed by space for the signature. The second phrase that may be used by a witness is "This work observed by me," or "Demonstrated to me," followed by space for the date and signature. This affirmation is used only when the witness actually observes an experiment, measurement, or other demonstration. Room should be available for the signatures of two witnesses. The subject of witnessing is discussed in detail from a legal standpoint in Chapter 7.

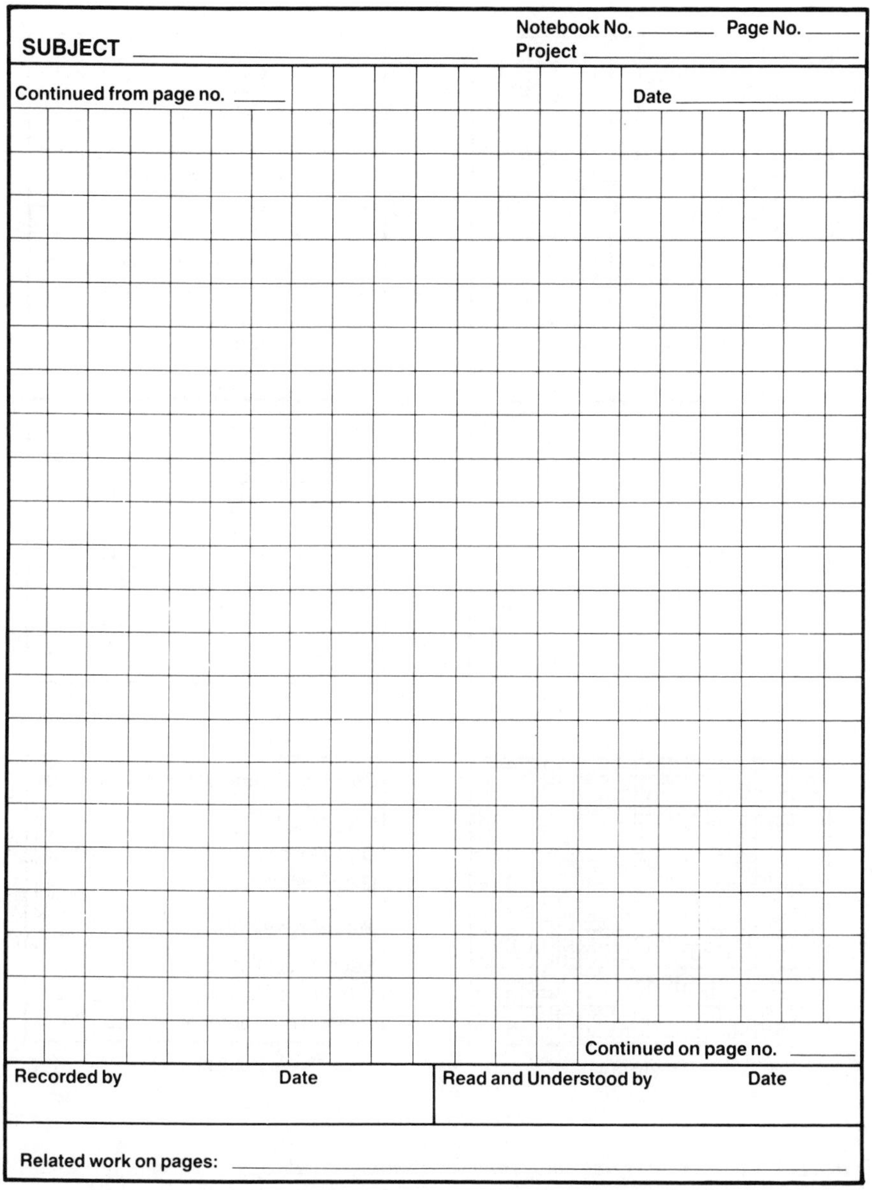

SUBJECT _____									Notebook No. _____ Page No. _____ Project _____						
Continued from page no. _____									Date _____						

Continued on page no. _____

Recorded by	Date	Read and Understood by	Date

Related work on pages: _____

Figure 2.3 An example of a blank notekeeping page. The narrow margins discourage marginal notes and provide space for clips to hold the pages during microfilming. The typeset words prompt the notekeeper to fill in essential information. See Chapter 6 for examples of complete notebook entries.

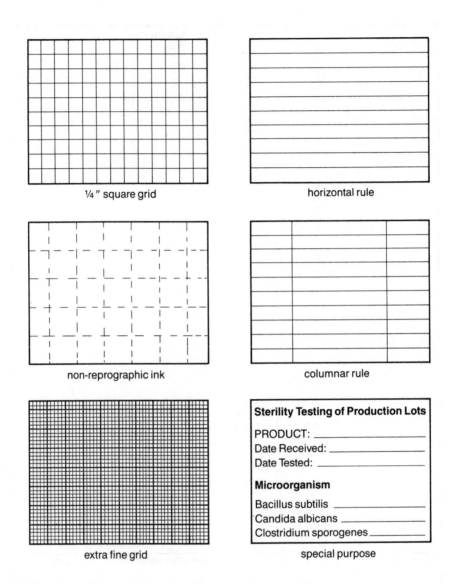

¼″ square grid

horizontal rule

non-reprographic ink

columnar rule

extra fine grid

special purpose

Sterility Testing of Production Lots

PRODUCT: _____

Date Received: _____

Date Tested: _____

Microorganism

Bacillus subtilis _____

Candida albicans _____

Clostridium sporogenes _____

Figure 2.4 Patterns of grids and lines commonly used on notekeeping pages. Many laboratories design pages for recording specific test results, as shown in the bottom right example.

Student Notebooks

Typical "composition notebooks" sold for student use contain horizontally ruled pages with no instructions or printed matter. They usually consist of a poor grade of paper and thin covers and are cheaper than the more professional type of notebook; 1984 prices are approximately $2 for the 100-sheet student "comp book" and $10 for the "professional" lab notebook. These cheaper notebooks may be fine for compositions, but they are not suitable for laboratory recordkeeping. Students of science, especially science majors, should get into the habit of keeping notes in a professional manner and using professional tools. The small difference in cost cannot justify using the cheaper substitute.

Notebooks with perforated, removable duplicate pages and carbon paper inserts are available and are especially recommended for students in laboratory classes and for young students undertaking their first research projects. The duplicate page feature is excellent for two reasons. First, the students are instructed to complete portions of the notebook during class and to hand in the duplicate pages before leaving the lab. This approach ensures that students learn to write down experimental procedures and observations as the work takes place. None should use the excuse, "I have to take the notebook home to write it over neatly." Second, the instructor will have a copy of the student's work that can be stored in a safe place in case the notebook is accidentally damaged or lost. To insist that duplicate notebook pages are handed in periodically is, in the long run, a courtesy and a help to the student, especially in high school and undergraduate research programs. Because the student investigator is inexperienced, the danger of losing or accidentally destroying notes is very real indeed. A small disadvantage of this type of notebook is that the original sheets must be rather thin (16-lb substance) to get a clear copy on the duplicate page. This lighter weight paper can be torn more easily than average notebook paper of 24- or 28-lb substance. Another disadvantage is that notes can only be written on one side of each page.

If the only notebooks available have blank ruled pages with no type matter, then the writer must adopt a consistent method of placing the page number, date, subject identification, and signatures. Getting into this habit will minimize the chance of forgetting to include these data at a critical juncture.

Photocopies

Nearly everyone in a research facility makes and uses photocopied documents. Plain paper photocopying is a cheap and fast method of reproducing memos, proposals, and reports. Articles published in technical journals are often photocopied so that they can be studied in detail, annotated, and dissected by investigators. Taping or pasting such photocopied material into the lab notebook for reference is common practice.

Photocopying is also used to preserve documents that are on unstable papers. An example of such use is related in the story of the American archaeologist who set out for several months of field exploration in southern Asia. He took with him several cases of high quality notepaper and notebooks. To his great disappointment, the paper was lost while on route and he was forced to buy locally available paper at the excavation site. Upon returning to his university in the United States, he had all the expedition's handwritten documents photocopied onto high-quality paper for long-term storage. Was this a wise decision? What is known about the permanence of photocopies?

The toner, which forms the finished image on plain paper photocopies, is a fine powder composed of carbon black in a thermoplastic carrier, usually an acrylate. The image is formed by electrostatic attraction of the toner onto the surface of a rotating drum. The toner is transferred to the surface of the paper, the paper is pressed against the hot rollers, the resin is melted, and the image is secured on the paper.

Because the toners contain carbon black, they are lightfast (i.e., stable toward fading in sunlight). The freshly formed image has abrasion resistance that is almost as good as ball-point pen inks. However, toner manufacturers have been unwilling to say that photocopies will have archival permanence, principally because the plastic carrier is susceptible to several means of attack. Plasticizers that are incorporated into some carriers can be expected to leach out of the image during long-term storage and perhaps embrittle the image and weaken the bond with the paper. The carriers can be attacked by several organic solvents, especially benzene, toluene, and acetone, but they have good resistance to water and alcohols.

Photocopies stored pressed against a sheet of plastic or synthetic fabric transfer their image to the fabric. Photocopies should only be stored between sheets of paper or other materials that do not contain plasticizers or leachable organic compounds.

To summarize, photocopies are relatively permanent if stored under the generally accepted archival conditions discussed in Chapter 4. Care should be taken if photocopies are to be attached to notebook pages: mending tape or adhesives must not be in direct contact with the printed image. Photocopying is a good method of preserving original documents that are known or suspected to be unstable, if the photocopies are made onto high quality paper.

Mending Tape and Glue

Often items are attached in the notebook such as photographs, stripcharts or printed sheets from analytical instruments, or even notes written on scraps of paper. For this purpose, mending tape or glue must be permanent, nonstaining, and nonreactive with the paper. Archivists and document conservators recommend starch paste because it meets the last two criteria. However, this paste is unsuitable for permanent research records because it can be dissolved and removed, an important consideration for the conservator but antithetical to the scientist's need for permanence. Starch paste must be cooked with water just before use, a messy and inconvenient process for the working scientist. The best alternative is to use a high-quality white glue that is an internally plasticized polyvinyl acetate emulsion. These products are permanent and acid free; they will not shorten the life of the paper, although they do tend to soak into paper fibers and cockle the paper. They are best suited for attaching relatively thick, nonporous items such as photographs to notebook pages. Use the paste sparingly and press the material flat while drying. Do not use rubber cement or epoxy adhesives on paper.

For attaching pieces of paper to the notebook pages, the best material is an archival quality mending tape manufactured in Great Britain and available through representatives in this country. (*See* the list at the end of this chapter.) It consists of a butylacrylate adhesive on an acid-free, long-fiber tissue paper backing. This tape is thin, nonstaining, nonyellowing, and close to neutral pH. In accelerated aging tests, it did not become brittle or react with the paper. For the best performance even with this tape, the tape must not be placed directly over handwritten or printed inks. Even this high-quality tape, as well as the adhesive previously discussed, can be dissolved with some organic solvents. The notebook must be protected from spills.

The common drug store variety of transparent mending tape is an inferior substitute for archival quality tape for two reasons. First, the adhesive can react with some inks and cause them to spread and fade. Second, the tape itself is not permanent and will eventually discolor, turn brittle, and release its hold on the document that it was intended to secure. The major American manufacturer of this type of tape disclaims archival permanence and will not specify how long the tape should be expected to do its job. The manufacturer will say that it should last longer than cellophane-backed transparent tapes, which were popular in the middle of this century and which generally lasted for only 10 years.

Summary

- Paper—Permanent paper for scientific notekeeping should contain 100% chemically purified wood pulp, have a minimum pH of 5.5, and contain no ground wood or lignin and no alum-rosin sizing. Durability is increased with long, strong wood fibers or cotton rag fibers. Additional permanence is achieved if approximately 3% calcium carbonate is present as an alkaline reserve.
- Ink—A black ball-point pen with a fine point is best. Some inks are more light-sensitive than others: all writing should be protected from sunlight exposure. Avoid colors, especially red, which is the most light-sensitive. Avoid porous, felt-tip pens; plastic-ball, roller-tip pens; and fountain pens.
- Notebooks—Case-bound, hard-cover notebooks with pages of permanent-durable paper are the best and the most expensive. Spiral-bound notebooks are not acceptable, and student composition books are an inferior substitute. Pages should be imprinted with page numbers and headings for vital information including subject, date, and author and witness signatures. Books containing 100 $8\frac{1}{2} \times 11$

in. sheets are generally the most useful, although other sizes are available. Books with duplicate pages are a must for teaching. Fit the notebook to the job.

• Photocopies—Unstable documents can be photocopied onto high quality paper for preservation. Dry toner photocopies should be placed between sheets of high-quality paper to avoid loss of the image by transference. The image must not be in contact with plastic sheets, synthetic fabrics, or mending tape.

• Tape and Glue—High-quality, acid-free white glue is the only type of glue recommended for attaching nonporous objects such as photographs to notebook pages. Use sparingly and press flat while drying to avoid cockling the paper. Do not use rubber cement or epoxy adhesives. Archival quality mending tape is best for attaching pieces of paper to notebook pages. Drug store variety transparent mending tape is an inferior substitute for archival quality tape.

Appendix 1. Manufacturers of Bound Laboratory Notebooks

The following two companies produce a wide variety of stationery supplies, including bound notebooks suitable for laboratory records. They sell large volumes in standard formats for mass markets such as universities. In addition, they manufacture custom notebooks designed to individual specifications.

Boorum and Pease
801 Newark Avenue
Elizabeth, NJ 07208
(201) 352-8800

Dennison National Company
(formerly National Blank Book Co.)
P. O. Box 791
Holyoke, MA 01041
(413) 539-9811

The following companies specialize in manufacturing bound notebooks for laboratory work. They supply many of the large industrial research facilities in this country and will accept small as well as large orders:

Scientific Bindery Productions
732 South Sherman Street
Chicago, IL 60605
(312) 939-3449

Scientific Notebook Company
P. O. Box 238
Stevensville, MI 49127
(616) 429-8285

Laboratory Notebook Company
P. O. Box 188
Holyoke, MA 01041-0188
(413) 532-6287

Cossart-Frederick Jones Co.
Division of Industrial Loose-Leaf
Corporation
31st & Jefferson St.
Philadelphia, PA 19121
(215) 684-1000

Eureka Blank Book Company
P. O. Box 150
Holyoke, MA 01041-0150
(413) 534-5671

In the summer of 1984, Texwipe Company announced a line of products made of nonparticulating writing materials, including lab notebooks, especially designed to be used in dust-free clean room environments. The address is P. O. Box 575, 650 E. Crescent Avenue, Upper Saddle River, NJ 07458, (201) 327-9100.

Appendix 2. Mending Tape and Glue

Archival quality mending tape is manufactured by Ademco Ltd., Coronation Road, Cressex Estate, High Wycomee, Buckinghamshire, England. It is available in the United States from Conservation Materials Ltd., 240 Freeport Blvd., Sparks, NV 89431; Talas, 213 W. 35th Street, New York, NY 10011; and University Products, P.O. Box 101, Holyoke, MA 01041.

Acid-free white glue sold under the trade name *Jade* is manufactured by Aabbitt Adhesives, Inc., 2403 Oakley Avenue, Chicago, IL 60647. The minimum order is 1 gallon for approximately $25. Smaller quantities are available throughout the country from conservation supply dealers such as Talas, Conservation Materials Ltd., and University Products.

Appendix 3. Pens and Inks

The nation's largest producer of ball-point pen inks also makes custom inks to order: Formulabs, P. O. Box 1056, Escondido, CA 92025. They formulated the Archival 101 ink, which is available in stock ball-point pens from Hub Pen Company, 26 Quincy Avenue, Braintree, MA 02184.

A permanent, black stamp-pad ink called *Manuscript Marking Ink* has been developed by the Library of Congress and the U.S. Government Printing Office (Formula No. 7837). Small quantities of the ink are available without charge from the Library of Congress Preservation Office. The ink is typically used to mark documents with ownership stamps and accession dates. The ink is claimed to have resisted 300 years of simulated sunlight and is inert to all known bleaching agents. The formulation is secret.

Appendix 4. Paper Testing Equipment

A spot-test kit for alum, lignin, and starch in paper is available for $25 from Applied Science Labs, Inc., 2216 Hull Street, Richmond, VA 23224.

A quick estimate of the acidity of paper can be made with the "pH Testing Pen" available from Light Impressions Corp., Box 3012, Rochester, NY 14614.

Literature Cited

1. "Paper and Its Preservation: Environmental Controls;" Library of Congress Preservation Leaflet, Oct. 1983, No. 2.
2. "Standard Specification for Bond and Ledger Papers for Permanent Records;" ASTM D3290–76; Philadelphia: American Society for Testing and Materials.

3. Kimberly, A. E.; Scribner, B. W. "Summary Report of National Bureau of Standards Research on Preservation of Records;" March 1937; NBS Misc. Publ. M144.
4. U.S. Government Federal Specification TT–I–563, Dec. 1930.
5. "Ball Point Pen, Single Cartridge;" U.S. Government Federal Specification GG–B–60D, May 1972.
6. "Ball Point Pen Refills: Quality, Specification, Testing, Coding;" DIN 16 554 part 2; Berlin: Deutsches Institut fur Normung.

Legal and Ethical Aspects

Ownership, Rights, and Obligations

Employees have well-established legal obligations concerning writings created by them that relate to their employer's business. The professional scientist also has rights regarding personal professional development and the ability to seek and to hold employment. The subject of this chapter is how these rights and obligations relate to notekeeping by scientists and technicians. Some of these rights and obligations have a legal basis, but much of a scientist's attitude about the thoroughness and care of recordkeeping has an ethical and professional basis.

In industrial situations, the employer owns the physical book and its contents; in universities the situation is less clear, and for that reason a brief discussion of notebook ownership is presented.

None of the information presented in this chapter should be construed as legal advice, but rather as a discussion of issues about which both employers and employees should be knowledgeable. A licensed attorney should be consulted for legal advice in any specific situation.

Employment Agreements

Both employers and employees stand to gain from good communication, and each should have a clear understanding of the other's position and expectations at the time of hiring. Most professional employees engaged in technical research and development, as well as technically oriented employees in other lines of work (e.g., production or market development), can be expected to have to sign employment agreements. Nonprofessional staff often do not sign employment agreements. The difference is based on the understanding that the professional employee works for the company 24 hours a day, 365 days a year and is being compensated accordingly. Any invention, development, or idea relevant to the business of the company belongs to the company, and the company has the right to select its line of business. The written employment agreement strengthens the company's position by calling the new employee's attention to this situation. Professional workers who do not sign explicit employment agreements are, in most if not all cases, bound by state laws on ownership of intellectual property such as patents and trade secrets. Most universities, on the other hand, encourage students and teachers to take with them the fruits of their labors and all of the knowledge gained while at the school.

The ACS Committee on Professional Relations has published a handbook entitled "Employment Agreements," which clearly and succinctly states the scope and purpose of such documents:

> ...[the employment agreement] is usually a brief (several paragraphs to several pages) legal document which describes what you will do for the company in return for your employment. Regardless of the length, it will generally consist of the following parts: (1) an agreement to assign patent rights, (2) a statement relating to safeguarding of trade secrets and other confidential information, and (3) a statement relating to the return of proprietary documents upon termination of employment.

The agreement is customarily signed by the new employee (and witnessed) on or before the employee's first day of work. (*Refer* to the ACS handbook for a detailed discussion of this material.)

Specific examples of employment agreement clauses that relate to recordkeeping are presented in the following sections.

Ownership

The notebook is the physical property of the employer in almost all instances and usually contains trade secrets or proprietary information in addition to general knowledge. Writings of professional full-time employees relating to the business of the company, which constitute most if not all of the notebook's contents, are the property of the company. The following are two examples of clauses from employment agreements, taken from the ACS handbook, that state this point:

> I acknowledge that all originals and copies of drawings, blueprints, manuals, reports, notebooks, notes, photographs and any other recorded, written or printed matter relating to research, manufacturing operations or business of [the company] made or received by me during my employment are the property of [the company]....

> All memoranda, notes, records, reports, photographs, drawings, plans, papers, or other documents made or compiled by or made available to EMPLOYEE during the course of employment with [the company], and any copies or abstracts thereof, whether or not they contain Confidential Information, are and shall be the property of [the company].

Most industrial companies have strict rules regarding the distribution and handling of notebooks and other records (*see* Chapter 4). At the other extreme, universities rarely have policies about the ownership or handling of notebooks. A survey of physical science departments at a cross-section of American universities was conducted in the preparation of this book. This survey indicated that it is a generally accepted policy for a student to take a copy of research notes written by the student while working under a professor's supervision. Professors often stated that they preferred to keep at least a copy, if not the original of the student's notebooks. Many professors stated that they could not possibly store all of the data generated by their students during the several years that the students spend pursuing an advanced degree.

Most large physical science departments allow each professor or individual research group to establish policy in this field. The universities have two concerns: First, they want professors to pursue research in the individual's chosen field with minimal restrictions. Second, most university departments do not want to get involved in administering centralized recordkeeping facilities. Therefore, professors may want to keep only the notebooks and thesis, but not other items such as the raw instrumental output, charts or graphs, and computer calculations. Of course, this attitude depends upon the type of research. In some disciplines (e.g., astronomy, geology, and archaelogy), raw data may yield useful results upon reexamination many years after the initial observations were made.

Students and professors should have a clear understanding at the beginning of the project about who is entitled to the original and copies of notes and data. If the work is industrially sponsored and may be classified as proprietary for some length of time, the student should be informed of this intention at the outset.

Some universities, following the example of industry, now require professors to sign employment agreements. In contrast to industrial agreements, these academic agreements usually provide for sharing between the individual and the university of income derived from the professor's inventions. As a result of long-term practice by universities, works of authorship, such as technical papers and books, are usually considered the property of the writer. This attitude may change as universities profit increasingly from high-technological inventions made by professors and students.

Employee's Obligations

In a recent survey of 81 companies, all of the employment agreements explicitly obligated the employee to disclose inventions that relate to the company's business and to help the company obtain patents. However, only 30% of the agreements explicitly mentioned the employee's recordkeeping duties. Two examples of such clauses are as follows:

> I will maintain accurate and complete written records and promptly, without request, disclose to [the company] all such inventions and discoveries made by me alone or jointly with others.

> I agree...To make and maintain adequate and current written records of all such inventions, discoveries and improvements, in the form of notes, sketches, drawings, or reports relating thereto, which records shall be and remain the property of and available to the company at all times.

Provisions of this type are included in employment agreements to ensure that documentation is available to support the company's patent claims. With millions of dollars often at stake in such proceedings, these provisions can raise some interesting questions: If the company loses a court battle because of poor recordkeeping by an employee, can the employee be disciplined or dismissed for failing to live up to the terms of the employment agreement? What constitutes adequate (good) versus inadequate (poor) recordkeeping? How does an employer know whether or not the employee is making a good faith effort to live up to the employment agreement provision?

Laboratory managers could spend some time thinking about how they would answer these questions. The answers, however, would only provide remedial action and might not repair some serious damage done to the company. Approaching the problem from another direction, the employer's time might be better spent by ensuring that new employees clearly understand what is expected of them in regard to recordkeeping duties. Toward this end, perhaps many more companies could have clauses similar to the two previously discussed, which explicitly state the employee's recordkeeping obligations.

Some workers think that they can preserve the security of their jobs by keeping certain essential information in their heads. They want to become indispensable to the company. For example, an employee might develop or improve a laboratory technique without ever writing the details into a notebook. The employee then thinks of herself or himself as indispensable, the only worker who can use the technique, and expects to be seen by the boss in the same light. Such an attitude is a disservice to both the employee and the employer and is a possible violation of explicit provisions in the employment agreement.

Supervisors and managers want to be sure that their operation does not become dependent upon any one person. In the event that the individual leaves the organization, or simply misses work because of an extended illness, other workers must be able to find out about the methods used by the individual and what had been accomplished up to the time the individual left.

From the employee's standpoint, the disservice done by not writing down certain notes is more subtle, and likely fateful for the employee's future. If the employee is truly indispensable, management will be unable to promote or transfer the employee unless the job knowledge can be passed on to someone else. Said simply, the wise employee knows how to be *valuable* without being *indispensable.*

Should some things not be written into the notebook? Up to this point, I have stressed the need to write down everything: Better to write all the details than to leave out a point that might someday be important. Should litigation occur, the chips will fall where they may. Another school of thought, held by some lawyers, states that information that might be found during the legal process called *discovery* and therefore useful to opposing parties in a legal action should not be committed to writing. If certain notes are not kept, such as the exact circumstances surrounding the conception of an invention, counsel can construct its case in the most favorable light. This view cannot be held by the average working scientist for the following simple reason: The good scientist does not trust to memory that which has been done, either in the lab or in the scientist's head. If a scientist is to be called to testify, the scientist must be able to look at notes to refresh memory and to state the facts precisely as they were observed.

Another reason for writing everything into the notebook is that scientists and supervisors do not know at the time of discovery whether an invention will be patented or practiced as a trade secret. If the invention is patented, documentation is essential to prove the date of conception and as a record of reduction to practice. If practiced as a trade secret, the only valid reason for not writing down the work would be to protect against espionage. Strict control of documents and reminding employees of their obligations are better alternatives. After all, the details of the secret might be lost or forgotten if not recorded. The unexpected departure of an employee, for a variety of reasons, is not uncommon. Therefore, employees should be educated and reminded about the importance of writing notes immediately after the discovery, invention, or observation has taken place.

Employer's Obligations

Besides paying a salary as compensation for the employee's work, are employers obligated legally in other areas related to notekeeping? Generally speaking, companies cannot unreasonably restrict the employee from taking general knowledge gained on the job to a new job. However, courts have upheld the employer's right to restrict the employee for a limited period of time—often 2–3 years—from going to work for a direct competitor. This restriction is based on the likelihood that the employee could not help but use at least some of the proprietary information gained at the previous employer's expense. The courts have generally required that the former employer pay the former employee's salary for that period. Also, where possible, the courts have allowed the employee to work for the competitor but have confined the employee to working in different areas of technology for the 2- to 3-year period (or whatever time period) of restriction.

The scientist's choice of career and job mobility is obviously complicated by these considerations. Companies therefore have the obligation to clearly inform employees about the employee's obligations as the company sees them. Even if the employment agreement spells out such obligations, the employee's supervisor should periodically review the situation to avoid misunderstanding.

The employer's obligation in terms of notekeeping is this: If and when the employee leaves, the employer is obligated to allow the employee to use general information acquired on the job. The employer is not obligated to let the employee take any physical property such as copies of journal articles, correspondence, or notebook pages. It is not unusual, but not necessarily typical, for an employee to leave a job empty-handed except for personal possessions.

Disagreement can arise concerning whether some particular portion of the notebook falls into the category of general knowledge or proprietary knowledge. Legal action to resolve the question is probably beyond the means of the employee.

Therefore, some amount of trust and mutual respect must be present in the employer–employee relationship. The employee should understand the company's position about removing information from the company's premises.

One ethical obligation of the company is to ensure that the employee is allowed to work in an atmosphere conducive to professional development and demeanor. Employees should be encouraged to keep good notes for the sake of their own professional development as well as for the good of the company. Projects should be planned so that employees can give care and thought to their work without constantly rushing to complete assignments. Time should be provided for the employees to peruse notes and to evaluate the direction of the work. They should not miss the forest for the trees.

Notekeeping is one of those general skills that is useful in almost any scientific career. It is surely a transportable skill and employees should continue to improve upon good notekeeping habits throughout their careers. Encouragement and constructive critical review of the employee's writing is another obligation of management.

Summary

The employee should

- Be aware of and live up to the explicit provisions of the employment agreement. Keep notes current and complete.

- Ask management permission before copying or removing any document from company premises, if there is the slightest possibility that the document contains proprietary information. Many companies have strict policies that forbid copying notebook pages without permission.

- Develop and improve notekeeping habits as a valuable personal skill.

- Approach notekeeping as a serious, integral part of the job, not a chore done in addition to the real work. Realize that much time, effort, and money can rest on the quality of recordkeeping and do as competent and professional a job as possible. See that the employer understands your needs for sufficient time to do an adequate job.

The employer should

- Clearly educate employees regarding recordkeeping responsibilities. Periodically review provisions in the employment agreement or other company policies relevant to this responsibility.

- Establish an atmosphere that encourages the thoughtful, professional attitude of employees. Do not expect everything to be done on a rush basis. Allow time for planning and reviewing work. Include time for notewriting when planning project schedules.

- Provide constructive criticism to improve the employee's writing skills. Encourage personal development of good notekeeping habits.

- When the employee leaves, have a clear understanding about what material, if any, the employee is permitted to take.

Management of Notekeeping

Practices for Issuance, Use, and Storage of Notebooks

To state that fame and fortune have been won or lost depending on the quality of scientific notekeeping is not an exaggeration. Remember the stories of LeMonnier and Gould from Chapter 1? Employers can adopt specific practices to increase the chances that their employees will keep complete and proper notes. No system is foolproof, but paying attention to certain principles can help to safeguard critical data and decrease costs. The most important principle is simply to provide clear, thorough instructions to new employees and to periodically reinforce these instructions.

In industrial settings, notebook contents are usually considered proprietary, and the worker has a responsibility to protect their secrecy. In academic and government laboratories, the information in the notebook will often end up in the public domain. In either situation, physical protection of the notebook is an important and often overlooked principle.

To determine the current practices of managing notebooks, a survey was conducted among some of the major research institutions in this country. Several universities, four major government laboratories, and a dozen industrial laboratories cooperated in the study. Most of the responding industrial organizations requested anonymity; therefore, responses are discussed without specific reference to their sources.

Phone contacts were usually made first, followed by correspondence and, occasionally, personal visits to the facilities. The people supplying the information were usually members of the legal staff, who reported official organization policy as they saw it. Direct contacts with researchers then revealed that these policies were indeed being followed to a high degree in nearly all of the organizations.

0906–5/85/0035/$06.00/1
©1985 American Chemical Society

Three questions were asked of each organization:
1. What instructions are given to employees upon hiring regarding the maintenance of research notebooks?
2. How is the distribution, collection, and control of notebooks handled?
3. Are notebooks reviewed on a regular basis as part of an employee's performance review?

In this chapter, the responses to these questions are discussed along with recommendations for handling notebooks. Good management practices that can serve as deterrents to dishonesty are described.

Instructions to Newly Hired Industrial Researchers

In industry, the amount of time spent by each organization instructing new employees on notekeeping habits varies from no formal instruction in some companies to lengthy orientation sessions conducted by others. These sessions are often led by corporate legal staff members and include an overview of the company's policies and philosophy toward inventions and discoveries, the basis of the company's success. Then, during the first week on the job, employees receive additional specific instruction on record keeping from the immediate supervisor. Many companies provide supervisors with detailed checklists of the subjects to be covered with each new employee. Some companies provide only general guidelines and leave the details to the discretion of the supervisors.

Interestingly, most companies provide these instruction sessions for professional employees and not for the technical support staff. At least one major company has the policy that only professional staff members can be issued serially numbered notebooks. Only in special circumstances when a need is demonstrated can a technician be issued a notebook. The point of such a policy is to ensure that only those workers who have been properly instructed about the legal ramifications of notekeeping should make notebook entries. Much of the value of the recorded information can be lost if notes are rife with speculation, incomplete entries, or improper witnessing. Such a policy is sensible for those companies whose incomes are based largely on patents, but it is of questionable value in other industrial situations.

Nearly all of the interviewed companies use custom printed notebooks that include one or two pages of explicit instructions for notekeeping. Several examples of such instructions are shown in Figures 4.1–4.3. During orientation sessions, instructors emphasize the following points:

• Notebooks are the physical property of the company and their contents are to be kept confidential. The worker who signs for and receives a notebook must safeguard it and see that it is returned for cataloging and storage after completion.

• Entries must be permanent, clear, complete, and made as soon as possible after the event (i.e., the experiment, observation, invention, or conversation) takes

LABORATORY NOTEBOOK

INSTRUCTIONS

This Laboratory Notebook is the property of [The Company]. It is assigned to you so that you may keep a complete, careful, chronological record of your work. The work which you do and the data which you enter in this book are confidential; they must not be disclosed to unauthorized persons. The Notebook must not be removed from the laboratory premises. Its preservation and maintenance are your responsibility; in case of damage, loss or disappearance, report the facts to your section manager at once.

The purpose of each entry in your Laboratory Notebook is to provide a complete record of your work, one that would enable one of your coworkers to repeat, if necessary, exactly what you did and secure exactly the same results, without having to ask any questions of anyone. You will find these specific instructions helpful in preparing entries which will meet this requirement.

1. Plan your experiment carefully, and plan the presentation which will best record the data you expect to secure. Since the duplicate page will be extracted from the Notebook and attached to the appropriate progress report for filing in the project file, data on each page must be limited to one specific project.

2. After Title and Project number, date and the objective have been filled in, your entry should record (1) the purpose of the experiment; (2) the materials used and their quantities; (3) the apparatus; (4) the procedure and manipulation (times, temperatures, pressures, pH's, and the like); and (5) the results. Where procedure or apparatus is standard, it is sufficient to describe it by reference; for example, ASTM D236-54T, or by reference to an earlier notebook page where it was fully described.

3. All data is to be recorded directly into the Notebook. Recording of original data on loose pieces of paper for later transcription into the Notebook is to be avoided. Should use of loose paper be necessary for proper conduct of an experiment, the loose paper should be signed and dated, and cemented into the Notebook. The data must be transcribed into the Notebook on the same day it was taken, and the Notebook entry should refer to and identify the loose paper which has been cemented into the book.

4. All entries must be made in ink. Erasures are not permitted. If a mistake is made, draw a line through the erroneous material and make a corrected entry immediately following.

5. Every entry must be dated, and signed at the foot and at the end of each day. In no event may the entries be signed less frequently than each page.

6. If, for clarity of presentation, it is desired to start the new entry on a new page when the previous page has not been entirely filled, draw a diagonal line across the unused portion of the page.

7. Pages must be used consecutively. Leaving a page or pages blank for later use is absolutely forbidden. Entries must be presented in chronological sequence.

8. If one of your coworkers (not a codiscoverer of the subject matter) has witnessed an experiment you have conducted, to an extent that enables him to state of his own knowledge what you did and what results you secured, have him sign and date the Notebook record of the experiment under the legend "Witnessed and understood by". If the experiment seems to you to be of sufficient importance, arrange to have it witnessed.

9. Pages are provided in the front of this book for an index to the subject matter covered. The index pages should be completed as the work progresses to afford ready access to the data recorded.

10. Avoid stating conclusions, particularly of failure or abandonment. Let the results speak for themselves.

11. When this book is filled, or upon termination of your employment, it must be turned in to your section manager.

Book No. _____ Assigned to _____ Date _____

Returned to Section Mgr.—Date _____

Figure 4.1 Instruction page from an industrial laboratory notebook.

Problem No. ....   Assigned to ....

Pages .... to ....

Date Started ....   Date Finished ....

INSTRUCTIONS

General:

This book is the property of [The Company]. Until it is returned to the Research Files its care and maintenance are the responsibility of the last person to whom it is assigned or charged. The contents will be microfilmed for permanent storage in the Research Files.

All entries should be in ink and dated, signed and witnessed in such a manner as to avoid any doubt as to when or by whom the entry was made. No erasures or eradications may be made. Make any necessary cancellations only by drawing a single inked line through the cancelled matter. Place the date and your initials at the end of the cancellation.

All notes and observations are to be entered directly herein, including tabulated data, calculations, drawings and graphical plots. When it is necessary to make records on separate sheets, these sheets are to be signed, dated and taped into this book. Predictions, comments, and interpretations of the observations may be important and should be entered. All such addenda must be attached to a blank notebook page to facilitate microfilming. Writing should be large and **legible.** The ink used must be blue or black and of such quality that it will not strike through to the reverse side of the page.

The first four pages are to be reserved for a chronological table of contents, which should be kept up to date while the book is still in use. The last few pages may be used for an alphabetical index.

It is necessary to fill each page and keep the sequence of entries in chronological order. Several pages may be reserved for a particular experiment, however. If the continuity of pages for a particular experiment is broken for lack of reserved space, notations should be made on both sides of the break. The unused balance of a page should be cancelled by a diagonal, inked line.

Each preparation, run, etc., should be entered on a new page and indexed. Such experiments shall be referred to by Problem and Page Number. Samples, material and equipment which are referred to should also be designated by Problem and Page Number, Pigment Code Number, or Plant Number where this will facilitate subsequent identification or reference. In general, where such items receive numbers in passing through our Manufacturing or Receiving Departments, those numbers should be retained through subsequent operations.

Records should be kept so as to permit the ready location of data and reference thereto, and must be intelligible to anyone who may be concerned with them later.

Ideas and Inventions:

Immediately upon conception of a new idea, record the idea on an Invention Record sheet with copies to Patent Law Department (2), Department File, co-inventor, and yourself. Send all copies to Patent Attorney's office for recording and numbering. Mere recording of the idea is not sufficient to establish inventorship and should be followed at the earliest opportunity by Reduction-to-Practice experiments performed in the actual presence of at least one technical witness who knows exactly what you are doing. The actual work outlined in steps 1 through 5 below, as well as written records of the same should be witnessed and dates recorded for each step.

Reduction-to-Practice Experiments:

Definition

A. One intended to give the first complete workable example.

B. The first experiment or series of experiments (in addition to that under A) intended to extend the idea beyond its original scope.

Procedure

1. Characterize all starting materials (old and new)

   a) By Source, Lot Number, date ordered and received.

   b) By Constants such as m.p., b.p., n^{25}_D, etc., plus characteristics such as color, odor, crystalline form, etc.

   c) By simple qualitative chemical tests or preparation of a derivative with a sharp m.p.

2. Actual Experimental Run(s)

   a) One Run or a series of runs on a test-tube or larger scale.

   b) Observe and record results in detail.

3. Characterize products fully.

4. Preserve a small sample (1 oz. bottle) for possible outside testing.

5. If the experiment provides the first workable example, fill out another invention record immediately and send all copies to the Patent Attorney's office. An Invention Record at this stage is known as a Disclosure Sheet.

Figure 4.2 Instruction page from an industrial laboratory notebook.

This Notebook Is Your Laboratory Diary

This notebook is the property of [The Company] and will contain confidential material. It should be treated as such and must not be taken from the Laboratory premises except in those cases where field work makes this necessary. This notebook is specifically assigned to you on a charge from the Library. It is your responsibility while it is in your hands. Upon being filled it must be returned to the Library. If needed again it should be obtained on a loan basis.

Original entries systematically and properly made in the usual course of your work are the best record of your accomplishments. This Notebook may be the deciding factor in litigation involving a question of inventorship. It may also be the only complete source of information for future experiments, or for the preparation of reports you may be required to write. THEREFORE:

1. As part of your daily routine enter clearly and concisely in this notebook all your experimental data, plans for experiments, analysis of results, calculations, observations or ideas. These data shall be placed in the notebook immediately and directly. Do not use loose leaf books or loose pieces of paper for this purpose. Make all entries in ink.

2. Each page and entry must show the date on which the entry is made and the signature (or initials) of the worker making the entry. The nature of the problem and the number of the project should also be indicated.

TO MAKE THIS BOOK MOST USEFUL YOU SHOULD:

(A) Plan the presentation before writing so it can be easily read and understood.

(B) State clearly and completely the procedure followed, giving all conditions of the experiment (temperatures, pressures, time required, etc.) and apparatus used (include sketches if necessary). This notebook should provide sufficient data to enable any of your associates to duplicate the procedure and results.

(C) Give all results and observations including references to literature, analytical data, and pertinent calculations. Data may be presented in tabular form and in graphs to great advantage. Cross sectional ruling is provided for your convenience in aligning tables and making graphs and sketches.

(D) Make all entries in ink. Always be neat and legible. Do not make erasures. Cross out by straight lines any material to be deleted.

(E) Make all entries consecutively, leaving no blank pages.

(F) When a page or a day's entry is completed, sign your name at lower left. If blank space is left on the page, draw a diagonal line to show its extent at the time of signing. Have your signature witnessed at the lower right. This witness means that the page was filled at the time of signing.

(G) Promptly disclose any new and possibly patentable ideas occurring thereon to one of your associates who is familiar with the work you are doing (preferably the section chief or one who has actually seen the experiment carried out), and have him affix his signature and the date. He should write in "disclosed to" at the time of signing.

(H) When an idea, process, etc., of any importance is finally found to be workable, the process, etc., should be demonstrated in the presence of witnesses, *other than co-inventors*, who are able to understand the process. These witnesses should sign the page in the notebook describing the demonstration carried out in their presence, writing in *"demonstrated to and understood by."* Before this procedure is followed, the section chief in charge of the project should be notified.

(I) Pages are provided for a Table of Contents. This should be completed to enable ready access to the contents in the future.

Book No. _____ Assigned to _____ Date _____

Returned to Library—Date _____

Figure 4.3 Instruction page from an industrial laboratory notebook.

place. Who did the work must be clearly indicated, and proper witnessing is essential. (More on this subject in Chapter 7.)

● The notebook is the place to record all ideas, communications, and work that relate to the business of the company, especially those subjects that might lead to new or improved products. Speculation is appropriate only when it concerns potential uses of an invention. In all other instances, stick to recording matters of fact.

In a nutshell, these are the principles of good notekeeping that should be presented to all new employees. In Chapter 6, these principles are demonstrated with examples of well-written notes.

During a symposium on laboratory recordkeeping held at the 111th ACS National Meeting in New York in 1947, essentially these same guidelines were presented as common practice in several companies. The only change since that time seems to be an increased emphasis on the need for careful witnessing. The principles for notekeeping that applied to that generation of research scientists are still applicable today.

Instructions at Government Laboratories

At government facilities, the attitude toward instruction is similar to that in industry, but perhaps a bit more relaxed. Usually each supervisor or group leader is responsible for seeing that new employees are properly instructed. The U.S. Department of Energy publishes a booklet entitled "You and the Patenting Process" (1), which is available to personnel at DOE laboratories and to grantees and contractors. This little booklet has good, brief instructions about notekeeping, including example notebook pages, and ought to accompany copies of the Uniform Reporting System for Federal Assistance that are given to government contractors and grantees.

Instructions Given to Student Researchers

At every university and college that responded to the survey, instruction on notekeeping depends on the attitude and training of individual instructors. None of the departments reported guidelines that were applied uniformly to all of their graduate students and faculty, although some research groups did have a consensus on how research notes should be kept. (In this chapter, I am referring to instructions given to student researchers and not to the general teaching of students, which is covered in Appendix A.) Some teachers, especially those with industrial backgrounds, are able to provide detailed instruction on notekeeping to students when they begin their research. Often, however, student researchers learn the art of notekeeping from a more senior student researcher who may or may not have had adequate instruction.

Several of the experienced industrial research supervisors complained that newly hired employees often lack adequate formal instruction in the principles of scientific notekeeping. The responses to the survey indicate that teachers seem to emphasize writing techniques that are used to prepare technical papers for publication instead of emphasizing the descriptive writing skills that are essential in an industrial career.

This impression was borne out during a seminar at the Eighth Biennial Conference on Chemical Education held at Storrs, Connecticut on 6 August 1984. Panelists and members of the audience discussed the topic "Writing Across the Chemistry Curriculum." Much of the discussion concerned how to teach the writing of technical papers for publication—generally, avoidance of the first person, active voice in the student's grammar (exactly the opposite of what is required of the industrial researcher). Most of the instructions given to industrial employees can be applied to student researchers.

In an environment of academic freedom, academic departments have a natural tendency to eschew prescribed discipline in notekeeping. However, more than two-thirds of the graduate students in chemistry (M.S. and Ph.D. candidates) will go on to work in the chemical industry. Notekeeping skills are indeed a valuable asset in an industrial career. Therefore, aren't department-wide, consistent, and uniform instructions appropriate for all student researchers? Notekeeping is a tool of scientific logic as well as a valuable writing skill. It is sometimes viewed, unfortunately, as an esthetic option rather than as a necessary scientific tool. Students should be encouraged and given the opportunity to develop the ability and the habit of good notekeeping at the same time that they are taught the theoretical principles of their science.

Issuance and Maintenance of Notebooks

Most corporations with significant research efforts supply their employees with custom printed, bound notebooks. Generally, one style of notebook purchased from a single source is used throughout the company, although many companies stock several sizes of notebooks useful for different purposes. The actual distribution of notebooks is often handled by one office at each research facility that has the responsibility to keep notebook issuance records. This function is usually carried out by an office of records management or by the research library. The distribution of notebooks is usually handled separately by each research center. This procedure makes sense because in large organizations the logistics of controlling the physical distribution of notebooks to many distant facilities is exceedingly difficult. On the other hand, some of the large companies do keep central records of the issuance of notebooks. For example, Exxon Research and Engineering has a comprehensive computer data system (2) that keeps track of every document (i.e., notebooks, reports, or memos) issued to any employee. Before the employee leaves Exxon, a list of all outstanding documents is given to the immediate supervisor with instructions to collect the documents before the employee leaves.

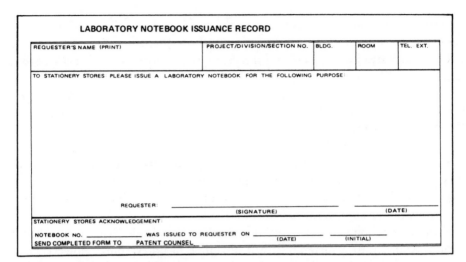

Figure 4.4 Laboratory notebook issuance record. These forms can be printed in carbon sets and bound into log books. Alternatively, this or a similar format can be set up as part of a complete database. (Reproduced with permission from Ref. 10. Copyright 1978, Prentice-Hall, Inc.)

Notebooks are almost always serially numbered and can be tracked even by a simple handwritten system. A form for this purpose is shown in Figure 4.4. This same type of form can be set up on a computer database system; an entry can be made when the notebook is signed out and again when it is returned.

Corporate research centers differ in their level of concern that notebooks be carefully protected while in use. Some companies have strict policies that notebooks be kept locked away whenever not in actual use. This policy is the best of all possible alternatives for several reasons. First, the notebook is protected from loss or destruction in case of fire or accident, and the opportunity for theft is reduced. Second, the notebook is always stored in a single location and can be immediately retrieved when needed. If the worker is not reminded to keep the notebook locked up, it will wind up on a desk one day, in a drawer another, and perhaps on an obscure shelf in someone else's lab or office at some other time. (Alas, even my own notebooks have occasionally been found in strange and unusual places.) If the notebook is kept locked up, the employee is encouraged to think of it as a valuable piece of company property rather than an expensive scratch pad. Notebooks are not permitted to be taken off the company's premises without authorization.

Photocopying of notebook pages is strictly limited by some companies as part of an attempt to protect the integrity of proprietary information.

Some industrial companies make a distinction between active and inactive notebooks. In general, *active* notebooks are those uncompleted books in which entries are still being made. If the notebook is filled to the last page, or if no entries have been made for a specified period of time, say, 2 years, then the notebook is classified *inactive* and must be returned for cataloging, microfilming, and storage.

In many laboratories, researchers keep their own notebooks in their offices for as long as they are employed. The argument made in favor of this procedure is that scientists often refer to notes made several months or years ago. Grabbing the old notebook off the shelf and looking up the needed information is easier than requesting official loan of the notebook from the records office or library and waiting for the notebook to be delivered. Keeping old notebooks handy makes sense for people in technical service and analytical laboratories who are often looking for methods or techniques that were developed once upon a time to solve a particular problem and that can be useful once again. Researchers in product development only need to keep on hand the notebook containing current work, including work going back perhaps as far as a year or two.

Many companies microfilm active notebooks to ensure that most of the contents are protected in case of loss and to provide an archival quality record of the notebook. Some companies perform microfilming only after the notebook has been filled and turned in for storage. The process of microfilming may be carried out by contract with an independent firm, although this method opens the possibility of disclosing confidential information to outsiders. Microfilming is done at least once a year for active notebooks.

The attitude toward issuance and maintenance of notebooks in government labs is similar to that described for industrial facilities. However, procedures such as the instruction of new employees are not as rigorously monitored. As in other areas, the attitude in government facilities appears to be somewhere between that practiced in industry and that practiced in academe.

In universities and colleges, students usually get their notebooks by requisition from the department stockroom or by buying them at the school bookstore. No universities reported having a central office responsible for the issuance and collection of notebooks used for research. Because funding for research is so diverse at even a single university, such a centralized system is probably not needed. However, someone, perhaps the department chairperson, should ensure that quality notebooks are available along with instructions for notekeeping. The distribution of notebooks to students might be handled differently from the purchase of other ordinary paper products. If the department or the research advisor provided the notebook directly to the student, its serious purpose would be evident.

Students will naturally take their notebooks in and out of laboratories, dormitories, and other places in the course of a day. Teachers should stress the value of the notebook and urge students to protect it.

Occasionally, student research notebooks contain proprietary information. Students working in universities under agreements with industrial companies should be clearly instructed about their obligations to grant or contract sponsors who require that information be kept confidential. Such instructions can make the student feel uncomfortable and it is up to the teacher to convey to the student the importance of honoring the sponsor's wishes. Students who graduate to become industrial researchers will have to sign employment agreements with secrecy provisions so they might as well learn to respect such provisions while in school.

Inspection of Notebooks While in Use

Of the dozen industrial companies surveyed, all indicated that supervisors do not monitor the notekeeping habits of their research employees as an official part of the employee performance review. Only one of the four government laboratories requires semiannual review of notebooks by supervisors. This finding was most surprising considering the emphasis that the companies put on notekeeping instruction and on protection of the notebook. These employers all recognize the value of proper notes, yet virtually none makes review of the notebooks a formal part of rating employee performance. The legal staff members and the lab managers who responded to the third survey question indicated that supervisors probably keep in touch with their employee's writings. How that monitoring was done, however, was up to each supervisor. In other words, some probably do and some probably don't.

How important is it that supervisors actually read their researcher's notebooks? Consider the following story about a research advisor who didn't.

A professor of chemistry at a major American university had a student who had obtained his doctorate degree and was ready to return to his native Japan. As is customary after completing the oral defense of the dissertation, the student and professor exchanged handshakes and polite wishes for future success. The student then presented the professor with his notebooks, samples, and other records, which the professor stored away for safekeeping. A few weeks later, the professor sat down and opened up one of the student's notebooks to verify some experimental work and was horrified to find that all the notes had been written in Japanese! He had not once looked at the student's notes during the several years that the work was in progress.

The supervisor who stays in touch with the researcher's notes will not be surprised when the time comes to review or repeat a critical piece of work. Young researchers, especially in industrial settings, should not feel insulted or belittled when occasionally asked to leave their notebooks for checking. After all, the industrial notebook is the property of the company and the worker has an obligation to follow the company's notekeeping policies. If such checking is emphasized when employees are new at their jobs, they will accept the procedure as an ordinary responsibility of employment. The supervisor can judge when the employee has developed the skill of notekeeping to an acceptable level. Past that point, checking of notes need only be done in connection with interpretation or validation of data.

The following is a ten-point checklist that supervisors can use when reviewing notebooks. All of these questions should be answered in the affirmative if the notebooks are being kept properly.

1. Are the notes written in black, ball-point pen?
2. Is the handwriting clearly legible? Are numbers and symbols unambiguous? (Does a *seven* look like a *two*, a *one* like an "el", or an "oh" like a *zero*?)

3. Is the table of contents up-to-date?
4. Is each entry signed and dated unambiguously? (Tues 9 October 1984, not 10-9-84)
5. Does each section have a clear, grammatical heading that describes the work reported therein?
6. Are entries written in the first person, telling who did the work?
7. Is the work described completely so that it can be understood without additional explanation by the writer?
8. Is the researcher "thinking in the notebook?" That is to say, are ideas and observations entered immediately and directly into the book and not on scraps of paper that are transcribed later?
9. Are entries witnessed correctly?
10. Is the notebook stored safely when not in use?

These questions cover only the mechanical aspects of proper notekeeping. The supervisor has a much broader responsibility to see that the researcher is "doing good science" by studying such factors as the reasonableness of hypotheses, appropriateness of controls, statistical analysis of data, or validity of conclusions. Many books detail the subject of managing scientific investigations, a few are listed at the end of this chapter. Further discussion of this topic is beyond the scope of this book.

These guidelines for checking notebooks are applicable in nearly every laboratory situation, yet each facility has its own special needs. Supervisors can spend a little time thinking about what aspects of their operations can benefit by special emphasis during instructions to researchers. Some examples are as follows: Are expiration dates of reagents noted? Is the last calibration date of the equipment recorded? Which spectrometer was actually used? Did the same technician prepare the samples each time? In the long run, to ask for detail is more cost efficient than to let assumptions take the place of facts. The motto ought to be, "Write it right the first time."

As for checking the notes of student researchers, I suggested in Chapter 2 that notebooks with perforated, duplicate pages were ideal for use by students. Research advisors can collect the notebook pages every few days and review the student's work at leisure. This procedure ensures that students are keeping their notes current and preserves a copy of the record in case the original notebook is lost or damaged by a lab accident.

Reducing Fraud in Research

Regular checking of raw data in notebooks and project files can catch sloppy or dishonest work before it makes headlines. In the late 1970s, several instances of fraud in both industrial and academic research institutes were directly traceable to lack of review of raw data by colleagues and supervisors. Scandals that drew national attention in the fields of biological and agricultural research occurred in large part because coauthors at distant research centers trusted their colleagues' claims of results but did not actually examine the raw data supposedly generated in the studies. Communication between coauthors was allegedly poor.

As a result of these incidents, several organizations developed guidelines to minimize the recurrence of dishonesty in research. The most detailed set of guidelines was published by the Association of American Universities Committee on the Integrity of Research (3). This group recognized that "Deviant actions by researchers may be grouped in four categories—scholarly fraud by falsification of data, plagiarism, abuse of confidentiality, and deliberate violations of regulations." To prevent dishonesty in research, the Committee recommended that attention should be given to six issues: Encouragement of intellectual honesty, discouragement of the attitude of "success at any cost," acceptance of responsibility by the laboratory director, maintenance of professional interpersonal relationships, establishment of well-defined experimental protocols, and appropriate assignment of credit and responsibility.

In that report, a specific guideline was stated for laboratory notebooks and other primary records:

> The director should supervise, teach, and encourage in-depth scrutiny and interpretation of results, emphasizing respect for primary data. Routine audit and review of all primary data by the laboratory director is strongly recommended. . . . Written, detailed, explicit procedures for data gathering, storage, and analysis are essential and should be available and practiced in all laboratories.

A three-member panel that investigated alleged fraud involving work funded by the National Institute of Health (NIH) at two major universities recommended seven steps to be taken at all NIH-supported clinical research centers (4). Among the guidelines was the statement: "Clinical studies performed by young investigators should be reviewed at regular intervals by the supervising physician, including raw data."

These guidelines place a heavy burden on the research director, who is usually too busy to check more than a small percentage of the total raw data generated in any given study. In many laboratories, the accuracy of raw data is customarily checked and the notebook entry initialed by a colleague of the researcher. This practice is common in pharmaceutical laboratories even though the U.S. Food and Drug Administration has stated (5) that raw data collected in the course of a nonclinical laboratory study does not have to be cosigned by a second individual. However, corrections to raw data must be accompanied by a signature and explanation of the reason for the alteration because the FDA believes (6), ". . . it is sometimes difficult for a second party, such as the personnel in your quality assurance unit, to distinguish 'obvious' errors [such as a simple incorrect date]. Consequently, the Agency insists that all corrections to raw data be justified." The accepted method of making such changes is by placing a single stroke through the unwanted entry, followed by dating and initialing the change with a word or two of explanation.

Collection of Completed Notebooks

Notebooks should be collected from their authors and put into permanent storage when employees (or research students) terminate their employment; work on a project has been completed and the project has been discontinued; or the notebook has lain idle for 2 years, even if the notebook is not filled to the last page.

These guidelines are subject to the previous exception: books to which frequent references are made can be kept handy as long as they are safely and securely stored under conditions that will minimize deterioration. At a minimum, notebooks should be kept in a closed cabinet rather than allowed to languish on a dusty shelf over bottles of chemicals.

When notebooks are collected for storage, the person responsible for receiving them should check for several items. Blank pages have to be marked off with an "X" and the signature of the author, or by a dated stamp that reads "No information on this page." The title page and table of contents can also be stamped with a phrase such as "This notebook complete on _____ (date)" or "Cataloged by Records Management Office. No entries after _____ (date)." The manuscript marking ink described in Chapter 2 is ideal for this purpose. Workers should make no further entries in books so marked.

The table of contents has to be complete when the notebook is turned in for storage. Many companies use the table of contents to create a computer-based index to the subjects contained in all research notebooks. This index is cross-referenced by author, project, subject and date and becomes an invaluable tool for determining work done in-house on a particular subject. Notebooks contain much more information than is ever put into technical papers or periodic internal reports; it is often difficult to determine from a report if a particular experiment was ever performed. The index can be rapidly searched to determine if the work was ever done, by whom, and when. Then, the notebook can be retrieved and studied in detail to determine the relevancy of the previous work to the current problem. Even desk-top microcomputers have sophisticated database management systems that can perform the indexing function for modest-sized laboratories. Once an individual leaves the facility, no one knows in detail all the little projects and experiments on which that individual worked. A comprehensive index of notebook entries can save much time and avoid needless speculation about the outcome of unpublished work.

Some organizations also require that the notebook author fill out a disclosure form that lists all the possible inventions described in that book.

After the completed notebook is accessioned into the collection, a microfilm or microfiche record is made. This photographic record serves two purposes. First, it is a copy that can be stored in a separate location to provide insurance against the loss of the data in the event of destruction of the original book. Second, the microfilm can be viewed when needed, and the notebook left protected in storage unless the original is absolutely required (e.g., to be introduced as evidence in litigation). This advantage reduces the handling of the original notebook and thereby prolongs its life.

A comprehensive book entitled *Legality of Microfilm* (7) covers the laws on admissibility of microforms as evidence in litigation. The laws of all 50 states are summarized along with policies and rules of major federal agencies. All of the relevant American National Standards Institute (ANSI) publications on micrographics, photographic film, and archival storage of film are also presented in that work.

In those laboratories that use notebooks containing duplicate pages, the researcher customarily turns in the carbon copies every few days. These copies may be stored by the supervisor, the records center, or the research library. The copies are kept until the notebook is completed; at that time, the notebook is archived and the duplicate pages are returned to the researcher or supervisor for handy reference. This system is ideal for student researchers because it allows them to keep a complete copy of their work while the school preserves the original.

Storage of Notebooks

What are the best conditions for storing paper documents such as laboratory notebooks? An experienced paper conservator once remarked that if paper records were kept in the controlled climate that we usually specify for our computers, the documents would still be in good condition long after the electronic equipment had become obsolete.

For over 50 years, archivists have recommended that paper records be stored at approximately 23 °C and 45–50% relative humidity. These recommendations have recently been confirmed in a Library of Congress publication (8). Cooler temperatures slow down the natural aging of paper: A rule of thumb is that the life of paper is doubled for every 5 °C reduction in storage temperature. Warm, moist conditions encourage the growth of molds and fungi that attack paper. At the other extreme, air that is too dry leads to embrittlement. If documents are to be retrieved several times each year, keep them at ordinary room temperature to avoid the mechanical stress that comes from the thermal cycling and attendant moisture condensation that occurs when they are taken out of storage. An interesting approach debated by some paper preservationists is that documents to be put away for long-term storage of many years can be sealed in moisture-proof plastic bags and stored at freezing temperatures. This approach is likely to be prohibitively expensive for large volumes of material.

Cool, long-term storage can be found in underground repositories, such as salt mines, in which space is leased expressly for document storage. Commercial warehouses are also available for document storage, if the research facility simply does not have sufficient room.

The 1937 National Bureau of Standards report on the preservation of records, which was mentioned in Chapter 2 in connection with inks, describes how the National Archives Building in Washington, DC, was constructed to create the best combination of conditions for paper document storage. Attention was paid not only to the control of temperature and humidity, but also to details such as alkaline wet scrubbing of the recirculated air to neutralize acidic pollutants and the use of specially selected structural materials to minimize airborne dust, which is abrasive to paper.

The facility chosen for storage of paper documents should be designed to minimize the risk of damage from fire or flood, to reduce the possibility of bacterial growth, to prevent encroachment of rodents and insects, and to be physically secure from theft or malicious mischief. Details of these and other considerations for a document storage facility can be found in "Quality Assurance Program Requirements for Nuclear Facilities," available from the American Society of Mechanical Engineers (9). (*See* Appendix 2.)

In universities, each department should evaluate its own needs for long-term storage of laboratory notebooks. Because these books are rarely, if ever, microfilmed, protection of the original is very important. A quick survey of how individual professors store their students' notebooks should be done to decide if changes are in order. Moving the notebooks from open shelves into file drawers or cabinets in rooms with temperature and humidity control will help to preserve the books.

Some departments currently insist that original notes stay at the university, although students are entitled to take copies with them. Retiring faculty at some schools are encouraged to make their own arrangements for disposition of their research data because no space is available to store such items. In rare instances, research notes are passed along to the library for archiving.

The driving force behind an effort to preserve academic research notebooks should be the recognition that a consistent policy for notekeeping will ensure continuity of the research record within each division or group. Incoming graduate students often read the thesis and notes of previous student research on a topic before actually beginning lab work. If each department prepares guidelines for recording and preserving original data, then each new student researcher can be sure of having previous notebooks available and in good condition.

After consultation with many academic department administrators, this recommendation is made with deep sensitivity to each faculty member's right to academic freedom and with full appreciation of the increased burden that will be placed on department officials. Once the guidelines have been developed and put into place, very little administrative effort will be required to see that notes are adequately preserved.

Preservation of Aging Notebooks

Countless notebooks made of cheap, unstable paper were used during the past 50 years in both industrial and academic environments. If the pages of important or especially useful notebooks are degrading as a result of mold, sunlight exposure, or aging, serious thought should be given to preserving them as soon as possible. The least expensive method is to photocopy the pages onto high quality, permanent-durable paper and to use these photocopies for referral. A second alternative would be to microfilm the notebooks. The third and most expensive alternative would be to send the notebooks to a qualified paper conservator for restoration. This restoration might involve cleaning, deacidification, and physical repair of damage. Such treatment can be expensive and should be considered for those notebooks judged to have potential historic significance.

Notebooks in need of conservation are probably too old to be of value for litigation and should be preserved either because they are of historic interest or because they contain useful, perhaps confidential, unpublished information. However, legal advice should be sought before original notebooks are disassembled for conservation.

All of these procedures will be in vain if the notebooks are not stored securely. Loss of notebooks, accidental or otherwise, is perhaps more likely in an academic setting than in industry because of the lack of controls on issuance and collection of notebooks in schools. The turn-of-the-century notebooks of Kamerlingh-Onnes containing the discovery of superconductivity are lost forever, but the notebooks of Michael Faraday and Alexander Fleming have been preserved. Clearly, the future of notes being written today is in the hands of research supervisors who can establish policy for their preservation.

Summary

- Instruction—New employees and student researchers should be given written guidelines that clearly state the organization's rules for notekeeping. Instructions based on the examples in Figures 4.1–4.3 can be made specific for any organization. Orientation sessions conducted by senior research personnel can convey the value of proper notekeeping to new employees.
- Issuance—Notebooks should be issued from a central office at each industrial facility, and records should indicate to whom the book was issued and for what purpose. At universities where student researchers buy their own books, professors should urge students to obtain high-quality notebooks with duplicate pages.
- Maintenance—Industrial notebooks should be kept locked in drawers or cabinets when not in use. Student researchers should be impressed with the need to protect the notebook from accident or loss. Duplicate pages should be turned in every few days. In industrial settings, the use of duplicate pages or periodic microfilming of active notebooks can ensure that data will not be lost.
- Checking—Supervisors should periodically check researcher's notes for compliance with the ten guidelines listed earlier in this chapter. Sloppy work and outright fraud can be virtually eliminated if checking by peers or supervisors is performed regularly.
- Storage—In industry, completed notebooks should be turned in to a central office for cataloging, microfilming, and storage. In academic settings,

duplicate pages given to the research advisor during the course of the work can be given back to the student in exchange for the original notebook, which will be archived. Conditions for storage should minimize deterioration of the paper and should protect the notebooks against accidental loss and theft.

Appendix 1. Reference Books

The following books are classical works on the methodology of scientific investigation.

Wilson, E. Bright *An Introduction to Scientific Research*; McGraw-Hill: New York, 1952.

Beveridge, W. I. B. *The Art of Scientific Investigation*; W. W. Norton: New York, 1951.

Bush, G. P.; Hattrey, L. H. *Teamwork in Research*; American University Press: Washington, DC, 1953.

Freedman, P. *The Principles of Scientific Research*; Pergamon: New York, 1960.

Appendix 2. Requirements for Records Storage Facilities

The following information is an excerpt from the "Quality Assurance Program Requirements for Nuclear Facilities," ANSI/ASME NQA–1–1983 Edition supplement 17S–1, p. 33.

Facility

Records shall be stored in facilities constructed and maintained in a manner which minimizes the risk of damage or destruction from the following:

    (a) natural disasters such as winds, floods, or fires;

    (b) environmental conditions such as high and low temperatures and humidity;

    (c) infestations of insects, mold, or rodents.

There are two satisfactory methods of providing storage facilities, single or dual.

Single Facility

Design and construction of a single record facility shall meet the criteria of (a) through (i) below:

    (a) reinforced concrete, concrete block, masonry, or equal construction;

    (b) floor and roof with drainage control. If a floor drain is provided, a check valve (or equal) shall be included;

    (c) doors, structure and frames, and hardware shall be designed to comply with the requirements of a minimum 2-hr fire rating;

    (d) sealant applied over walls as a moisture or condensation barrier;

(e) surface sealant on floor providing a hard wear surface to minimize concrete dusting;

(f) foundation sealant and provisions for drainage;

(g) forced air circulation with filter system;

(h) fire protection system;

(i) only those penetrations used exclusively for fire protection, communication, lighting, or temperature–humidity control are allowed; all such penetrations shall be sealed or dampered to comply with the minimum 2-hr fire protection rating.

The construction details shall be reviewed for adequacy of protection of contents by a person who is competent in the technical field of fire protection and fire extinguishing. If the facility is located within a building or structure, the environment and construction of that building can provide a portion or all of these criteria.

Literature Cited

1. "You and the Patenting Process;" U.S. Dept. Energy, 1980, DOE–GC–003. (Available from U.S. Department of Energy, Assistant General Counsel for Patents, Washington, DC 20545.)

2. Landsberg, M. K. "Security and Information Transfer;" presented at the 188th National Meeting of the American Chemical Society, Philadelphia, August, 1984.

3. Danforth, W. H.; Arnott, S.; Barber, A. A.; Ferrendelli, J. A.; Ross, R. S.; Sachs, H. G.; Travis, C. K. "Report of the Association of American Universities Committee on the Integrity of Research", 1983.

4. Norman, C. *Science* 1984, *224(4649)*, 581.

5. Lepore, P. D. "Good Laboratory Practice Regulations—Questions and Answers;" 1981, p. 22. (Available from Bioresearch Monitoring Staff, HFC–30, FDA, 5600 Fishers Lane, Rockville, MD 20857.)

6. Lepore, P. D. "Management Briefings on the Good Laboratory Practice Regulations—Post Conference Report;" 1981, p. 35. (Available from Bioresearch Monitoring Staff, HFC–30, FDA, 5600 Fishers Lane, Rockville, MD 20857.)

7. Williams, R. F., Ed. *Legality of Microfilm*; Cohasset Associates: Chicago, 1980. (Available from the publisher at 3806 Lake Point Tower, 505 North Lake Shore Drive, Chicago, IL 60611.)

8. "Paper and Its Preservation: Environmental Controls;" Library of Congress Preservation Leaflet, Oct 1983, No. 2.

9. "Quality Assurance Program Requirements for Nuclear Facilities;" ANSI/ASME NQA–1–1983 Edition supplement 17S–1, section 4.

10. Carlsen, R. D.; McHugh, J. F.; "Handbook of Research and Development Forms and Format"; Prentice-Hall: Englewood Cliffs, New Jersey, 1978.

Organizing and Writing the Notebook

Be Flexible

The key to writing a useful notebook is simple clarity: Clear layout, clear descriptions, and good penmanship. A notebook that is a continuum of scribbles and scrawls from cover to cover will waste your time when you try to locate old information and may actually be misleading. The easier it is to find and understand subjects written into the notebook, the more use you will make of it. No single system for setting up the notebook will be the best system for all people in all situations. Each job, each project, each person needs a system that is right for the moment. You must adopt a flexible attitude about how to set up the notebook. In this chapter, guidelines are presented for organizing your writing for different purposes. Keep in mind that anyone should be able to pick up your notebook and understand what you have just written.

You need a flexible attitude because you must consider your own perspective when you write the notebook as well as the perspective of future readers. If you are recording analytical test results, a straightforward tabulation is most sensible because you and future readers can best make sense of carefully laid-out data. Another phase of a project, such as running a prototype research apparatus, is best handled by a narrative description, as if you were telling a story to someone who has not seen the apparatus. Flexibility in organizing the notebook means sometimes abbreviating subjects and sometimes spelling them out in detail.

A speaker's maxim is to know the audience to be addressed and to suit the talk to the audience. So it should be in writing the notebook. As you change jobs, you will change your perspective: sometimes you will be writing notes strictly for yourself, sometimes for a boss, and sometimes for subordinates. For these other people to understand your writing, you have to put yourself in their position and realize what they know or don't know about the work. You may also want to

0906–5/85/0053/$07.75/1
©1985 American Chemical Society

consider that some work will be of value to an unknown person who, years later, may pick up your notebook in an attempt to find a fact or two.

The contents of any notebook are broadly divided between the front matter and the body of the book.

Front Matter

The front matter of a laboratory or research notebook comprises the following elements: the title on the spine or cover, the signout or issuance page, the page of instructions, the table of contents, the preface, and the table of abbreviations.

Exterior Title

Consider how helpful the title of the notebook would be on the spine or cover so that it can be read in a stack of books on a shelf. Because most notebooks have dark covers, you can use white typewriter correction fluid to paint a title vertically on the spine of the notebook. For the title, you may use a project number, project name, or simply a sequential number. Titles should be simple, such as "NEW PIGMENTS," "HPLC DEVEL," or "MICROS." Scientists who are not issued serially numbered notebooks often use a code consisting of their initials followed by a sequential number for each book. My first notebook would be labelled "HMK01," the second "HMK02," and so forth. This simple sort of coding is valuable for identifying samples and for tying computer or instrumental output to the correct notebook.

Signout Page

The signout page was discussed in Chapters 2 and 4, and an example of a signout page was shown in Figure 2.1. If a printed signout page is not present, you can easily copy by hand the form shown in Figure 4.4.

Printed Instruction Pages

The instruction pages were discussed in Chapter 4, and several examples were shown in Figures 4.1–4.3. Each company, school, or research group will have its own variation of an instruction page, but they all should have essentially the same guidelines.

Table of Contents

The table of contents usually follows the instruction page. It can be very handy for quickly finding specific topics in the notebook but it must be kept up-to-date to be useful. The table of contents should have just enough information to be useful but not so much information that it is hard to read. It can be organized in several ways depending on the work reported in the notebook (*see* Figures 5.1–5.3). A general purpose table of contents can include one column each for the date,

TABLE OF CONTENTS — *Analyt. Chem 233-a*		Notebook No. *HMK-1*
Date	**Subject**	**Page No.**
9 Sept. 1984	Preface	1
(Begun 10 Sept. 84)	Table of abbreviations	2
10 Sept. 1984	Determination of chloride by gravimetry	3
17 Sept. 1984	Equivalent wt. of a solid acid	7
1 Oct. 1984	Detm'n of oxalate by $KMnO_4$ titrimetry	10
8 Oct. 1984	Fe in ore by dichromate titration	14
15 Oct. 1984	The titration curve and ionization	—
—	constants of phosphoric acid	20
5 Nov. 1984	Potentiometry with quinhydrone electrode	26
12 Nov. 1984	Formation constant of $Ag-NH_3$ complex	32
19 Nov. 1984	Solubility of $AgCrO_4$ by conductometry	39
4 Dec. 1984	Controlled-potential coulometry	44
10 Dec. 1984	Summary of Experimental Techniques	
	I Learned this Semester	52

Figure 5.1 Table of contents from a student's laboratory notebook.

TABLE OF CONTENTS – *TECH. SERV. DEPT.*			Notebook No. CP-1123	
Date	Project No.	Subject	Client	Page No.
6 Jan. 1985	TS-01185	Discolored coating	Dept. 62	28
8 Jan. '85	TS-01385	Brittle polypro. liner	C.G. Harris	31
26 Jan. 1985	TS-01485	Black specks inside new vinyl cases	Sales Dept.	40
4 Feb. 1985	TS-01885	Calibrate microscope	—	46
6 Feb. 1985	TS-02285	Identify fibers	Process Eng.	47
8 Feb 1985	—	Repairs to DSC	—	49
14 Feb. 1985	TS-03185	Holes in Al-Ti ribbon	Don Hall	52
17 Feb. 1985	TS-03485	Discolored coating	Dept. 62	58
2 March 1985	TS-01885	Calibrate microscope	—	74
3 March 1985	TS-03985	Scratches in Hi-density	—	—
—	—	polyethylene bottles	Sales	75
9 March 1985	TS-03485	Discolored coating	Dept. 62	79
17 March 1985	TS-04485	"Sparkling Powder"	Alice Lowe	84
22 March 1985	TS-04985	Floaters in CHCl$_3$	HPLC Lab.	89
26 March 1985	TS-05085	Install new ion	—	92
—	—	chromatograph	—	—
6 April 1985	TS-01885	Calibr. microscope	—	115
4 May 1985	TS-01885	calibr. microscope	—	116
6 May 1985	TS-8485	Identity fibers	Process Eng.	117
9 May 1985	TS-08685	Unknown dust found	—	
—	—	on filter belt	Q.A. Dept.	121
12 May 1985	TS-09485	Tears in PMP tape	Sales	127
16 May 1985	TS-09985	"Seeds" in bottles	J.K. Moyer	131
21 May 1985	TS-10485	Green spots/stripes	Proc. Eng.	134
30 May 1985	TS-03485	Discol. coating solved!	Dept. 62	144

Figure 5.2 Table of contents from a notebook used to record technical service investigations. Entering the project number, subject, and client name will aid in future searching.

TABLE OF CONTENTS — Quality Control			Notebook No. QCD-11425	
Date			Subject	Page No.
Run Began	Product	Reactor	Comments	—
0845 hr. 1 June	KT-22	Kettle #6	NORMAL	2
1015 14 June '85	LC-21-1	Extr. 14	<u>Low steam - long warmup</u>	4 ←
0900 17 June '85	KT-22	Kettle #6	Dble. batch	8
0730 19 June '85	KT-22a	Kettle #4	New stirring blade	12
0800 24 June '85	KT-22	Kettle #4	Replaced thermocouples	13
0715 9 July 1985	KT-24	Kettle #6	Contaminated! Discarded	17 ←
0715 11 July '85	KT-24	Kettle #4	NORMAL	23
0730 14 July 1985	KT-24	Kettle #4	NORMAL	27
0845 17 July 1985	KT-24	Kettle #4	Valve stuck - overheat	34
0800 10 August	KT-22a	Green	NORMAL	39
0745 17 August	KT-22a	Kettle #6	Windows etched !?	46
19 August '85	REPAIRS	TO KETTLE #6		49
0830 24 Aug.	KT-22a	Green	NORMAL	56
0745 26 Aug.	KT-22a	Kettle #6	Normal - Blue cast?	59
0810 27 Aug	LC-21-1	Extr. 11	Replaced dies	67
0830 29 Aug.	KT-24	KETTLE #4	NORMAL	69
1 Sept. —	VACATION	SHUT DOWN & MAINTENANCE -		74

Figure 5.3 Table of contents from a notebook used to record quality control work. Note the use of arrows to indicate unusual observations.

subject, and page number. If many projects are recorded in the notebook, such as in technical service or analytical work, the table of contents can have a column for the project code number. If the notebook is used for a long-term research project, daily or weekly summaries of progress can be indexed in the table of contents. If the notebook is used to record quality control data, the table of contents can be set up to list lot or batch numbers, or times and dates when raw material sources are changed.

The philosophy of flexibility is obviously at work here. Don't stick to writing one style of table of contents for your entire career; set up each table of contents so that you will get the most from it.

Preface

Probably the most neglected part of notebook writing is the preface. When beginning a new notebook, take a few minutes to write a couple of paragraphs that describe the author and the purpose of the notebook. Although you know the purpose of your notebook, future readers may be stymied without a brief preface. The preface need not be longer than one page; you should include the following information:

1. Who is the author? Give job title, department, and specialization. Who is the supervisor? List coworkers who are contributing to the project.
2. What is the goal of the work? How far along was the project when this book was started? Summarize the progress to date.
3. Where is the work being performed? This point can be useful for a future reader who wants to find associated equipment or records. Where are other records and samples stored?
4. Who is funding or sponsoring the work? Who is the contact person for the funding? Include the address and telephone number for handy reference.

Two examples of a preface are shown in Figures 5.4 and 5.5. One illustrates a typical student preface, which might be written at the start of thesis research. The other shows a preface that might be written by an industrial researcher at the start of a product development project.

Table of Abbreviations

The last item of front matter is a table of abbreviations, symbols, code numbers, or other information that is used throughout the book. Some people put such information on the inside front or back cover, a good idea because information can be easily found there. An example of this sort of table is shown in Figure 5.6. Again, be flexible: If you don't think that such a table is useful, then don't include

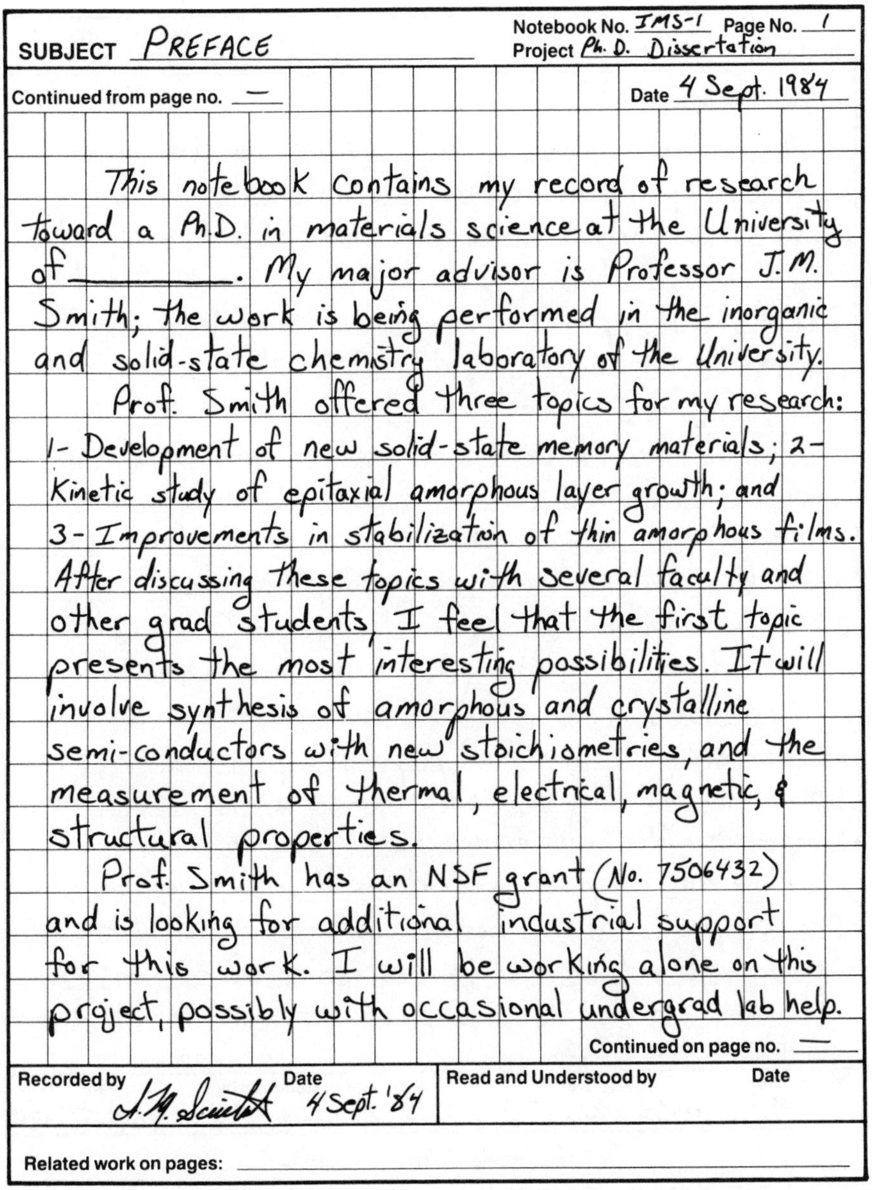

| SUBJECT _PREFACE_ | Notebook No. _IMS-1_ Page No. _1_ |
| | Project _Ph. D. Dissertation_ |

Continued from page no. ____—____   Date _4 Sept. 1984_

     This notebook contains my record of research toward a Ph.D. in materials science at the University of _____. My major advisor is Professor J. M. Smith; the work is being performed in the inorganic and solid-state chemistry laboratory of the University.

     Prof. Smith offered three topics for my research: 1 - Development of new solid-state memory materials; 2 - Kinetic study of epitaxial amorphous layer growth; and 3 - Improvements in stabilization of thin amorphous films. After discussing these topics with several faculty and other grad students, I feel that the first topic presents the most interesting possibilities. It will involve synthesis of amorphous and crystalline semi-conductors with new stoichiometries, and the measurement of thermal, electrical, magnetic, & structural properties.

     Prof. Smith has an NSF grant (No. 7506432) and is looking for additional industrial support for this work. I will be working alone on this project, possibly with occasional undergrad lab help.

Continued on page no. ____—____

| Recorded by _d. M. Scientist_   Date _4 Sept. '84_ | Read and Understood by   Date |

Related work on pages: _____

Figure 5.4 Preface to a notebook used by a student to record dissertation research.

SUBJECT _PREFACE_	Notebook No. CX-2421 Page No. 1
	Project _low-cap. ion exch. substrates_

Continued from page no. —   Date _10 Oct. 1984_

   This notebook contains Michael J. Edwards' record of work on novel, low-capacity ion exchange substrate materials. I am an associate chemist in the new products department, part of the Retail Products Div.; the work is a team effort, led by Dr. J. S. Thompson. Others on the team include Meg Jones (polymer characterization), Fred Davis (synthesis), and Alicia Freidkin (kinetics & mechanisms).

   We will attempt to develop materials to replace aging product lines-- new, improved versions of our old reliable polymer beads such as EXCEL-122 and ACEL-14. Marketing Dept. has told us that we can displace competing products if ours have better resistance to low pH (2-2.5) and less tendency to cracking when thermally shocked.

   Up to this point, Dr. Thompson has reviewed our analyses of competitor's products and prepared a plan of attack. It is up to me and Meg to assess the behavior of Fred's new compounds. We heard at the last Pittsburgh Conference that several companies are interested in such new materials. One company, "ION-X," has already signed a confidentiality agreement and agreed to test some of our materials.

   Samples and records of completed work are being kept in Bldg 226 - Rm B348.

Continued on page no. 2

Recorded by	Date	Read and Understood by	Date
M. J. Edwards	10 Oct. 1984	_L. Ossembrana_	10 Oct. 1984

Related work on pages: _Dr. Thompson's notes in notebook CX-2404_

Figure 5.5 Preface to a notebook used to record a long-term industrial research project.

SUBJECT *ABBREVIATIONS*	Notebook No. *CX-2421* Page No. *5*
	Project *Ion exch. beads*

Continued from page no. ——   Date *17 Oct. 1984*

The following abbreviations are used often in this notebook:

Abbrev.	Meaning
NP/RPD	New Products Dept. of Retail Products Div.
TMD	Technical Marketing Dept.

Instruments:

FTIR-1	Nicolet 5DXB spectrometer, tag #17-044 Rm B11
IC-1	DIONEX 2010i, serial no. 1742, Rm B12
IC-2	Waters ILC-1 ion chromatograph, Rm B12
DTA	DuPont 900 thermal analyzer
Micr-1	Leitz polarizing light microscope in petro lab.

Samples will be coded like this:

CX2421-5-a

notebook number   page no.   sequence on page

Continued on page no. _____

| Recorded by *M. J. Edwards* | Date *17 Oct. 1984* | Read and Understood by *L. Quesambrana* | Date *17 Oct. 1984* |

Related work on pages: *other abbreviations on back inside cover.*

Figure 5.6 This table of abbreviations is extremely useful because it permits the use of shorthand notations throughout the body of the notebook. Some of this information can be filled in as the notebook is being written. However, the table should be complete when the notebook is turned in for storage so that future readers will understand otherwise cryptic references.

it. At least take a minute to think whether including such a table will make your work a little easier, not to mention improving clarity for other readers.

Some researchers like to put index tabs at the beginning of major sections of the notebook. For example, tabs on the first page of the table of contents and at the table of abbreviations can be handy time-savers.

Numbering the Pages

Some research notebooks available at universities and nearly all industrial notebooks come with printed page numbers. If you have to use a notebook without printed page numbers, take a few minutes and number the first dozen or so pages when you sit down to write the preface. Put the page numbers in the upper outside corner of every page, starting from the very front of the book where the preface or table of contents will begin. Write the number legibly and enclose it in a circle so that it will not be mistaken as part of the date, data, or a doodle. Always keep a few pages numbered ahead of where you are writing your notes so that you can reference work on one page to work on another. Don't fall into the trap of writing a few pages without paying attention to page numbers and then forget to fill in the table of contents or make cross-references.

Summary of Front Matter

When you get a new notebook, you will need less than 30 minutes to label the spine, write the preface, start the table of abbreviations, and begin numbering the pages. You and your readers will find this preparation well worth the effort.

Remember, too, that the notebook is a tool, not an end in itself. Don't make notekeeping the *goal*, but rather one of the *means* of achieving your particular scientific goal. A well-kept notebook is a pleasure to read and a valuable asset, but don't spend excess time perfecting the appearance of your notes at the expense of work on the actual scientific problem.

Some Simple Practical Suggestions

Make writing space when setting up experimental apparatus on the lab bench. Put the notebook on the bench and *then* set up the apparatus. This sequence forces you to create plenty of room to work and write. Don't try to squeeze the notebook into a small space between pieces of equipment. You will likely leave a corner of the notebook sticking out over the edge of the bench just waiting for someone to knock the book (and perhaps the apparatus) onto the floor. If you are right-handed, put the notebook on the right side of the equipment on the bench; if you are left-handed, put the notebook on the left side of the equipment. Some old-style lab benches have pull-out "courtesy shelves" that are ideal for notebook writing.

Don't use the notebook as a tray to carry small objects such as crucibles, sample boats, or glassware. I have seen several days' work crash to the floor when a technician who was balancing several samples on a notebook rounded a hallway corner and bumped into another worker going in the opposite direction. Carry the notebook under your arm and carry samples and apparatus on trays or carts with ample sidewalls.

The Body of the Notebook

During the course of your career, you will use notebooks for many activities: to plan work and experiments; make safety notes; record data and observations; perform calculations; discuss results of work and experiments; describe inventions; track personnel assignments; review progress; and keep track of reagents, supplies, and equipment performance.

The concept of flexibility can also be applied to the body of the notebook. Think of each job or project that you are about to begin recording as a separate entity from previous entries, and ask yourself, "How should I organize the headings for this project's topics for maximum clarity?" Routine work often needs no more than a sentence or two of background explanation before getting into the actual experiment, observation, or measurement. Long projects, on the other hand, can definitely benefit from thoughtful choice of notebook headings and organization.

Each job or project can usually be broken down into four major phases: *background, planning, execution,* and *study of results.* You can describe each of these phases by using one or more section headings in the notebook. You can vary the amount of space devoted to each of these sections depending on your need to give either a brief summary or a detailed description of the work.

Laboratory Research Experiments

In the typical laboratory or pilot scale experiment that might last from a few hours to a few weeks, the following section headings should be used: "Introduction," "Experimental Plan," "Observations and Data," "Discussion of Results," and "Conclusion."

The first three headings correspond to the first three major phases listed previously. The phase called "Study of Results" has been divided into two sections, "Discussion of Results" and "Conclusion." This division allows you to use the discussion section to "think in the notebook" about what you have done. The conclusion section then follows naturally and should be a brief, clear statement of your accomplishment. Note that these section headings parallel the scientific method of deductive reasoning, which encompasses creating a hypothesis, devising an experiment to test the hypothesis, conducting the experiment, and drawing conclusions (a step that often means revising the hypothesis).

For the researcher who works more often by inductive reasoning (gathering many facts and then drawing a conclusion), these headings are still satisfactory. Begin by stating the broad goal of the investigation in the introductory section. For each separate test, write an experimental plan followed by the observations and calculations pertinent to that test. Postpone the conclusion section until you have sufficient data to make a guess. These five "tried-and-true" headings will suffice in almost all experimental situations.

For each section, you may use general headings, such as "Introduction" or "Experimental Section," or you may use specific headings relevant to the particular project, such as "Background and Need for a New Photochromic Polymer" or "Experimental Plan to Test Stability of Dye Compound #X21776." Specific headings are preferable. If a single section runs many pages in length, you can use subtopics of that section as additional headings; write them at the top of each page.

In university research laboratories, one student usually performs most, if not all, of these steps and keeps the complete record in a single handwritten notebook. Such a notebook should be very easy to follow because section headings are in sequence; each experiment is clearly separated from the next.

In industrial laboratories, however, some of the work will be performed by people other than the notebook's primary author. The notebook will not always follow the simple order of section headings. For example, some work will be recorded on printed forms called *bench sheets*. Some of the work, such as analytical or characterization experiments, might be done by technicians in other departments who provide reports in the form of computer printouts. You, as author and experimenter, might write the description of some experimental work performed today, but you might have to wait several weeks to get test results from another lab before you can write the discussion and conclusion sections. For that reason, you must give each section of the notebook a clear heading and include appropriate references to page numbers of relevant sections. For example, when freshly synthesized samples are sent out for analysis, note on the current page that they were delivered to the analytical lab and when results are expected to come back. When the results are received, copy them into the notebook and refer to the page on which the synthesis was described. As another example, you should describe a newly constructed apparatus only once and, in future use of the apparatus, reference the page on which the initial description was made.

INTRODUCTION   Begin recording an experiment by starting a new page with the date, project number, title, or other specific appropriate heading. Immediately below that information, write the introduction, a clear statement of the scientific problem. Some researchers use the heading "Purpose" rather than "Introduction" because this section focuses on the specific short-term goal of the work. As in any good piece of writing, begin the introductory paragraph with a topic sentence that states the purpose of the work. In the remainder of the introduction, provide explanation and support of the proposed work: Why is the work being undertaken? What related work has been done (by others, yourself)? Cite the literature. What were the results of this previous work? Why was the current experiment chosen? Show the relevant chemical reactions and other calculations that were needed to plan the experiment. What will be the benefits if the experiment "works"?

For routine work, the length of the introduction may be as brief as a single sentence: "Today I received two samples of paper (nos. 25671 and 25688) from Dr. Morgan in Dept. 233 to be tested for pH in accordance with TAPPI method 435." On the other hand, the introduction to a lengthy research project may warrant many pages and consist of separate subsections with headings such as "Literature Survey," "Calculations and Reaction Pathways," "Possible Experiments," or "Predicted Properties of the New Compound." During the course of experimentation or data analysis, you will frequently refer to the introductory section to check calculations or to reassess the reasons for choosing a particular approach to the problem. Therefore, after several pages of writing, summarize your work up to that point and clearly mark the summary section.

THE EXPERIMENTAL PLAN   A description of the planned experiment should follow the introductory section. If you omit the experimental plan and jump into a narrative description of the actual observations and data collection, then your readers will likely have trouble understanding what you were doing. A clear statement of the problem means that you understand what you want to do and that you are focused on a particular approach to solving the problem. For large projects, a well-described plan is a help when you begin to make flowcharts of tasks. When assigning work to others, a clear statement of the plan is essential so that you and your coworkers are on the same wavelength. Because laboratory experiments are often modified while in progress, the statement of the experimental plan preserves a record of your original intent. This section may help avoid unwittingly repeating previous work; therefore, include it in every write-up.

Use simple sentences to state the work to be done. You can clarify your planned procedure by drawing a flowchart, outline, or numbered list of the experimental steps. State whether the work will be done by you or by others.

References to previous work can often substitute for writing detailed experimental plans, but should not substitute for careful experimental planning. One caution should be mentioned here: Do not confuse this aspect of notekeeping with the research leader's need to write detailed experimental plans. Many types of research, especially work in the health, pharmaceutical, and biological fields, must have detailed experimental protocols. Government agencies that sponsor such research generally do not require that protocols be handwritten in the lab notebook. Indeed, such protocols often run to many dozens of typewritten pages and are not appropriate for the notebook.

Information presented in the introduction should not be recapitulated in the experimental plan section. The introduction should be a general look *backward* at what has been done to date and the reasons for undertaking the current problem. The experimental plan should be a look *forward*, describing the specific work to be done. If the plan is presented as a list or outline, you can insert the appropriate page numbers when each item of the plan is actually accomplished.

Include notes on safety in this section, along with the properties of the materials that will be used and precautions for the various operations to be performed. A table of the properties of the materials you plan to use will be helpful. Such a table may include chemical formula, formula weight, melting or boiling points, color, viscosity, specific gravity, vapor pressure, and other useful properties.

THE OBSERVATIONS AND DATA SECTION   This section might be considered the heart of your notewriting because, in it, you actually record the observations that you make during the course of an experiment. These notes and data will lead you to accept or abandon a hypothesis and help you decide the course of future experiments. You must be as objective and honest in recording your observations as you are in making them.

In this section, you will record raw data (also called primary data): the actual measurements of variables such as mass, length, intensity, and time. This raw data is precious. Treat it with the care and respect that you would give to a family heirloom. Indeed, the data may be passed on to another generation of scientists who have to trust your observations and make use of your work.

Record the data as completely as possible and leave interpretation of the observations for later. With this attitude, you are better able to concentrate on your experimental observations. You will be less likely to make mistakes if you are not at the same time trying to perform calculations *and* draw conclusions. To get the most from your work, wait until the experiment is complete before deciding which data are useful, which data are useless, and which data need additional study. This approach implies that you should be thoroughly familiar with the experimental plan and that you anticipate possible outcomes, even if you favor only one result. In other words, to paraphrase a famous baseball player, the experiment isn't over until it's over.

However, you should always be on the lookout for the novel, unexpected happening. As Pasteur observed, "Chance favors only the prepared mind." Some of the most remarkable discoveries in history were based on chance observations, such as Fleming's discovery of the action of penicillin on bacteria, von Roentgen's discovery of X-rays, and Gram's discovery of the differentiating stain for bacteria, which now bears his name. When you make an unexpected observation, be sure that you have all the experimental details noted or reproduction of the conditions that led to your discovery may be impossible.

Be sure that your penmanship is at its best when recording your observations because an illegible datum may require repeating some of the work.

Most of the observations and data section can be a narrative description, a story, telling what you did and what you saw. Use the first person, if appropriate, to make clear that you did the work. If someone else did the work, be sure that point is obvious. (Questions of inventorship may have to be resolved on the basis of your descriptions of who did what, and when.) Write in reasonably brief, declarative sentences as the work progresses. If you write convoluted or complex sentences with many parenthetic phrases, perhaps with chains of dependent clauses (not a good idea anyway), you will make future reading difficult and may very well obscure the main points of your work.

General Principles of Recording Data   More than sixty years ago, Reginald Hughes wrote, "The cardinal merits of good notes are lucidity, brevity, and simplicity." Today we say, "Notes should be clear, concise, and complete." To write the most useful notes, you must follow several principles. The most important principle is that observations and data must be recorded immediately.

You have two good reasons for the immediate recording of raw data, one obvious and one subtle. Most importantly, the immediate recording of raw data serves to preserve the research record. If you trust to memory any observations with the expectation of writing them into the notebook later, you will surely forget something. For example, don't try to remember the weight of a sample. If the phone rings, or a colleague interrupts, you may easily forget what might be an important number that you had hoped to put into the notebook. The second reason for immediate recording of data has to do with keeping your mind clear and alert while the experiment is in progress. If you begin to clutter your brain with minutiae, you will be unable to think clearly about the work at hand. Use the notebook as a tool to record what you have done or seen and then move on to the next part of the work.

Simply put, write down whatever happens, when it happens. Don't wait until the end of the day to sit down and recollect your thoughts; you must plan for adequate time to write notes. Don't try to squeeze your notewriting into spare moments between the more exciting aspects of experimentation. Make notekeeping, like safe working habits, an integral part of whatever you do.

A researcher's demeanor generally falls between two extremes. Some are methodical and concerned about every step they take, Pasteur-types (P-types), and some rush headlong to the lab, eager "to get their hands dirty" without adequate planning, LeMonnier-types (L-types). L-types keep sloppy and mostly useless notes, whereas P-types keep the best and most valuable notes. Undoubtedly, you should have some characteristics of both types within you and bring to the surface either the enthusiasm or the caution, as the situation warrants. Overall, probably the best attitude is one weighted heavily in favor of careful planning, with restrained eagerness. If you are the type of person who has trouble sitting down and writing detailed notes or if you like to try out every idea with "just one quick experiment," you will have to work especially hard to develop the habit of patient, complete recording of your work. The job will be easier if you have a supervisor who has and who expects you to have such an attitude.

You must employ the doctrine of flexibility when planning how to record your raw data. Formats suitable for recording various types of experimental work are limited only by your imagination. (Several specific examples are given in the next chapter.) Any format you select must meet the general criteria for logical order and legibility. The set-up of data tables should lend itself to ease of interpretation and should minimize the chances for transcription errors. For example, if you are recording data that requires some calculation followed by plotting (either by hand or by computer), set up your data table so that the sought-after X and Y coordinates will be in the last two columns or rows. This arrangement will make the plotting easier with less chance of transposing numbers. Make tables with plenty of room to record the data. Ask a colleague to check that your original data were transcribed correctly. (Some research protocols require that all data be checked and initialed by a second or even a third person.)

Page Formats   Some people advocate using right-hand pages for the body of your text and using left-hand pages for calculations, drawings, or other miscellaneous activities. From a legal consideration with an eye toward proving priority of invention and diligence in reduction to practice, you should use every page in sequence without ever leaving blank pages. (More on this subject in Chapter 7.) Some notebooks have right pages imprinted with horizontal lines for writing text and left pages imprinted with grids for graphing or making mechanical drawings. If you are uncomfortable writing text on the gridded pages, simply cross out the entire left page, sign it, and date it with a phrase such as "No entry on this page." A rubber stamp is convenient for this purpose and can also be used to mark off unused pages at the end of the notebook.

Writing the Date   The date should always be written near the outside margin of each page, that is, ending at the right edge of right pages or beginning at the left edge of left pages. This habit will help you to find a specific dated entry when flipping through the notebook some weeks or months after the entry was made. For example, if you want to review a piece of work that you recall doing in the springtime 2 years ago (or was it 3 years ago), you will easily find the entry if you always write the date in the same way.

Several formats are commonly used to write dates, but I recommend using this format: 25 April 1983. Other formats in use include: (American) April 25, 1983 or 4/25/83, and (European) 25/4/83, 25.4.83, and even 25.IV.83. The rationale for the last example is that the Roman numeral is an unambiguous indicator of the month and is independent of the language in which the notebook is written. Obviously, 4/25/83 and 25/4/83 are unambiguous dates (unless they are sample code numbers!) compared to 2/1/83, which could mean the second of January or the first of February. Computers often are programmed to print the date in a condensed form, such as 25APR84, that is also acceptable for handwritten notes.

Many researchers have the bad habit of omitting the year in the date entry. I have come across entire notebooks in which no year is to be found! Put in the complete, unambiguous date rather than leave out part of it. Putting in the day of the week is a matter of personal choice; it can be omitted with little concern. However, if you accomplish a particularly significant piece of work you may want to include a brief phrase such as, "29 Nov 1985, Friday after Thanksgiving." This little note may serve to jog your memory at some point in the future and may help to emphasize the certainty of when the work was performed.

PAGE NUMBERS   Notebook pages should be serially numbered in the upper outside corner. Take care not to obliterate the page number with an insert or with your writing. If your notebook does not have printed page numbers, write the page number by hand and enclose it in a circle to set it off from the data. Alternatively, use some notation, like "p. 29," to make clear that it is a page number and not part of your data or observations.

Pages must not be torn out or added to the notebook for any reason. Such action opens the possibility of questioning the authenticity of data.

Getting the Details   How much detail should be recorded in your notes? Could another scientist who is competent in your field pick up your notebook and repeat your work solely from the written description without additional explanation? If the answer is yes, then you are doing a good job. Too many details are better than for you to assume that a future reader (perhaps you!) will know all about your work. A story told by Nobel Laureate, Sir Geoffrey Wilkinson, illustrates this point:

Mr. A. J. Shortland made the first synthesis of hexamethyltungsten, one of the biggest breakthroughs in transition metal organometallic chemistry, by a method involving interaction of methyllithium with WCl_6 in ether. This was described in *J. Chem. Soc. Dalton Trans.* 1973, 872, as well as in his Ph.D. thesis in which we had stressed the necessity for rigorous oxygen-free conditions.

It came to my notice subsequently that some others had been unable to reproduce the work. Consequently I had a new student, Mr. L. Galyer, try to repeat the work as described and he also failed. Since Dr. Shortland now had a teaching job at Dulwich College in London, I asked him to come back and demonstrate—which he willingly did. The crux was that he had degassed the petroleum used to extract the evaporated reaction mixture separately in another flask by bubbling nitrogen through it. Instead of transferring the petroleum to the reaction flask via a steel tube and serum cap technique, he actually removed a stopper and poured the petroleum into the flask. This immediately partially resaturated the solvent with oxygen—irreproducibly of course depending on the pouring path length, flow of nitrogen out of the joint from the reaction, etc.

We are pretty sure that the initial reaction excess $LiMe + WCl_6$ gave a lithium alkylate anion of the type $Li_2[W(IV)Me_6]$ with reduced tungsten and that oxygen was required to reoxidise this to WMe_6.

We had a similar experience with synthesis of $ReOMe_4$ from $ReCl_5$ (*J. Chem. Soc. Dalton Trans.* 1975, 607) in which we utilized this observation.

If it had not been for Shortland's sloppy technique we would doubtless have abandoned work on high oxidation state methyls.

The point of this little story is simply that no experimental detail is too insignificant to be omitted from your notes. Procedures that have become routine present the insidious problem of being so habitual that you forget they might not be known to other people. One way to check that you have recorded all of the procedure is to ask a colleague who is familiar with your work to read your experimental description. Often what is obvious to the writer will not be obvious to the reader. In university settings where graduate students are typically working alone and instructors may have many students each working on a different problem, the instructor may not be able to review the student's notes to check for completeness. Each student must be sure that the notes being written are indeed comprehensive.

Don't be embarrassed to describe what you think is a sloppy or poor technique. Your results hinge on what you actually do, not on how nicely you tell everyone it was done.

In the chemical research laboratory, pay attention to the following details:

● Record all information necessary to unambiguously identify chemical reagents and other research materials, including the source (manufacturer), lot number, purity, type of container, and age or expiration date.

● Note whether water was distilled (singly or multiply) or deionized. Was it stored before use? What is your check on its purity?

● When using an instrument, note the last calibration date. How do you know it is performing properly? When was the last time that oven or furnace temperature sensors were checked?

● Use proper names for labware and vessels. Was the sample weighed into a boat, dish, crucible, beaker, or flask? What kind of flask?

● Record the materials that vessels are made of: Are your crucibles platinum, gold, ceramic, graphite, or something else? What procedures were used to clean and prepare glassware, mixers, reactors, or other vessels?

● In what sequence were the reagents mixed? Was "A" added to "B" or vice versa? How precisely were reagents measured? Did you use a pipet, buret, graduated cylinder, or automatic dispenser? Was the balance readable to 0.01 gram or 0.00001 gram?

● How were materials heated? What was the heating rate? How were materials stirred? Gently or vigorously? Mechanically or by hand?

● Was the laboratory atmosphere unusually dry or humid, hot or cold, dusty or smelly? What caused the unusual condition?

● If you are recording the color of a material, are you observing it under fluorescent, incandescent, or natural light?

● How long did it take to go from step A to step B? Did the precipitate appear immediately or after an hour? At what point did you stop working? Did the experiment sit unattended over the weekend?

Researchers working in pilot-scale plants and other industrial process operations should pay attention to the following subjects: It is not sufficient to precisely record a temperature; you must also indicate how and exactly where in the set-up the temperature was measured or recorded. This variable is very important in catalysis when temperature-sensitive reactions are being investigated. Often the recorded temperatures will appear in patent claims or publications and can either discredit the author or provide a basis for avoidance of infringement.

Flow rates in continuous reactors can be reported in many ways. For example, the volume of liquid per internal volume of the reaction zone may be used, or the volume of a gas at standard temperature and pressure per volume of free space in the reaction zone may be used. Mass velocities can be based on the mass of a catalyst component, including the support. Definition of the units employed is obviously very important when space or mass velocities are reported. Sufficient information must also be included so that conversions to other commonly used units can be made.

Yields and reports on conversions per pass present an even greater number of alternatives and options to the notebook writer as well as a chance for confusion and error at a later time. These observations can be based on material charged or material converted, or material converted to the highest value product either produced or recovered in pure form. The term *ultimate yield* implies recycle to extinction of a reactant and you should indicate whether or not a purge stream was involved. Many hours of high-priced professional time has been spent haggling over such differences and omissions in vain attempts to make meaningful deductions or comparisons.

Reporting *acid strengths* involves a wide choice of terminology to the keeper of the notebook. A lack of uniformity and understanding has led to endless confusion and argument among lawyers and evaluators. Normality, pH, or grams per liter have different values depending upon the methods and the internal standards used.

Many methods have been developed and used to estimate the average molecular weight of polymers. Reporting the degree of crystallinity, density, and viscosity of polymers presents similar problems. These methods have undergone evolutionary changes over a half century. Careful identification of the method used and definition of the units is very important in recording average molecular weights.

Drawings   A good drawing can save you several pages of writing and often conveys a better sense of the device you want to describe than words alone could do. Mechanical devices, novel labware, special combinations of existing equipment, and electronic circuit diagrams are often the subject of such drawings, whereas common lab apparatus need not be illustrated.

When making a drawing, keep the following points in mind: Note whether or not the drawing is to scale. What is the scale? What are the materials of construction? How are joints or connections made? If you invent a small part that fits into an existing apparatus, make an overall drawing that shows how the whole thing fits together. If the device has moving parts, indicate the possible directions of motion. If material will be moving through the apparatus, indicate the directions of flow. Make the drawing big enough to allow labels and comments to be written alongside appropriate parts. Use shading or cross-hatching to show features rather than using different color inks.

Don't spend extra time making the drawings pretty; they should be simple and to the point. Ask a colleague to look over your completed drawing to see if it makes sense or if it requires clarification.

Graphs   In this age of computers, plotting data by hand is becoming increasingly rare. Whether you plot data yourself or with the aid of an electromechanical contraption, the following features belong on every one of your completed graphs:

1. Put a title and date at the top of the page or immediately under the graph (like a newspaper caption).
2. Label the axes. Put in tick marks and dimensions on the axes.
3. Define the plotting symbols in a caption or legend.
4. Include error bars if you know the error of your measurements.
5. Generally, plot no more than three or four curves on the same graph. Keep it readable.
6. Use different types of lines (i.e., solid, dashed, or alternate dot-and-dash) instead of using different colors.
7. Note on or near the graph where the data that were used to plot the points can be found.

Attaching Loose Sheets   Chapter 2 contains a brief discussion of how to attach items such as instrumental output and photographs to notebook pages. What kinds of items should actually be attached to the notebook and what kinds of items should be kept separately in file folders? Generally speaking, charts and photographs should not be attached to the notebook unless they are needed to prove an extraordinary point. The accumulation of extra documents between notebook pages will break the binding of the notebook. A rule of thumb to follow is that any piece of paper larger than a single notebook page ought to be stored separately, correctly cross-referenced to the notebook page. Correct cross-referencing means that the original output should be dated, signed, and witnessed (if appropriate), and the relevant notebook page should be noted. If the original data are securely stored, the notebook only needs to contain a summary or condensed form of the findings. For example, the original paper stripchart (or computer hard copy) of an absorption spectrum should be catalogued and stored, but the notebook entry only needs to state the absorbance at the wavelength of particular interest.

If instruments are printing the results of tests, be sure to record in the notebook pertinent information such as the machine's serial or identification number and its most recent calibration date. Record on the instrumental output (i.e., stripchart or X–Y plot) all of the instrument's settings, along with the project and sample identification.

Instrument charts slightly wider than a single notebook page can be attached to a notebook page and folded to lie across the opposite page when the notebook is laid flat. Under no circumstances should an insert extend beyond an outside edge of a page because it will not fall within the page area delineated for microfilming and the chart edges will become ripped and frayed by ordinary handling of the notebook.

Inserting Specimens Work such as quality control or development of coatings, fabrics, or films sometimes requires storing samples in the sequence they were produced, or adjacent to each other for reference and comparison. Special notebooks can be purchased that have extra space in the binding to accommodate the inserted samples. (*See* the note on this subject in Chapter 2.)

Code Numbers for Specimens How should you assign code numbers to samples or to other written records? Probably the best system is to use your notebook number followed by the page number and if necessary, a sequential digit or letter. Thus, a strip chart or a sample could be labelled HMK02–76a, where the notebook is number HMK02, the relevant information is to be found on page 76, and this reference is the first for that page. Because industrial lab notebooks are sequentially numbered, this coding scheme produces a unique identification number (e.g., 11038–76a) that will not be accidentally assigned by another researcher to another sample.

Many companies do have formal sample numbering systems, based on production lot numbers, department numbers, or simply on centrally issued, sequential numbers. With the trend toward computer-generated instrumental analyses, many reports are automatically assigned numbers by the computers. If your company or department does not have a formal system for assigning numbers to samples and instrumental output, check what other people have been doing and select a system by consensus. That way, your colleagues, subordinates, and superiors will all be able to understand notations in each other's notebooks and other documents.

Some companies encourage their scientists to put notebook page references into reports. Although inclusion of a notebook page number is inappropriate in a paper intended for publication in the refereed literature, notebook page references should be inserted in analytical lab reports, project memoranda, or periodic research progress reports. *Drafts* of published papers can be annotated with notebook references and filed in the event that correspondents want additional information about published work.

Continuation of Work If your writing is a continuation of an experiment recorded on previous pages, precede your current entry with the notation, "continued from page ___." When you come to a stopping point and you expect to continue writing several pages later on, make the notation, "continued on page ___." If you stop while in the course of a chemical reaction or recrystallization, note clearly at what exact point in the procedure you stopped and for how long.

Signing Entries At the end of each work session or when you reach the end of a page, sign the page and date your signature. Have a witness also sign the page. This aspect is especially important if the work may result in a patentable invention. (See Chapter 7 for more on witnessing.) If you complete your record of work and do not wish to use the entire page before beginning another page, then strike a line or "X" through the blank, unused portion of the page. This mark indicates that you intentionally left that portion of the page unused. Do not leave whole pages blank with the intent of going back and filling in data or discussions at some later time.

Making Corrections   Make corrections to data or calculations by crossing out the incorrect data with a single stroke, accompanied by your initials and a brief explanation of the reason for the error. Be sure to leave the unwanted entry legible; it may turn out to have been correct after all! If you want to cross out an entire row, column, or page of data, put a circle around such unwanted data along with your initials and a note of explanation. If you perform an experiment and discover that something went wrong early in the procedure (such as starting with the wrong solvent or an incorrect mass of material), do not delete the data. Instead, mark at the top of the page, or in some other prominent spot, the fact that an error was made and that the correct data can be found on another page. Many experiments will not produce the desired results, but the data generated and the knowledge gained are usually just as valuable as if the experiment had succeeded. Nobel Laureate Sir Peter Medawar wrote, in "Advice to a Young Scientist," "If an experiment does not hold out the possibility of causing one to revise one's views, it is hard to see why it should be done at all." If you obliterate the data in a fit of anger, you may lose something valuable.

DISCUSSION OF RESULTS   Begin the discussion section with a heading such as, "Discussion," "Interpretation," "Evaluation of Data," or some other phrase that clearly separates this section from the data and observations.

This section provides you with the opportunity to reflect on what you did and what you saw during the course of the experiment. Write this section after the observations are completed. If an idea occurs to you while recording the observations and data, put it into the notebook at that time to preserve continuity and document the circumstances under which an invention or discovery was made. However, as a matter of general practice, concentrate on making observations while performing the experiment and concentrate on interpreting the data after they have been recorded. This section can contain calculations, charts or graphs, tables of rearranged or interpreted data, and prose ramblings. It is the section in which you truly "think in the notebook." An idea may occur once, and only briefly: catch it while you can and freeze it onto the page for further study.

Often you will want the opinion of a colleague regarding some recent work and your interpretation of the data. Many researchers find it a pleasant courtesy if you hand them a notebook with several pages marked and ask them to take a look at your writing, rather than try to lecture them about your observations and interpretation. If you take the time to thoroughly study the data, you might see more than your original interpretation. A thorough discussion section is better than a brief summary of what you found. Do not use the discussion section to simply restate the data. Use it instead to *understand* the data. If the data clearly fit your hypothesis, then say so. If the two are at odds, say so and discuss why. Speculation is appropriate in this section, but not in the section used to record the observations.

CONCLUSION   In this last section of your notes, you should summarize the goal of your work, what was done, and what you found. Typically, the conclusion can be handled in one or two pages for long projects and in just one or two sentences for routine work. The page on which the conclusion appears can be indexed in the table of contents or marked with index tabs for ready reference. Several subjects should be included in the conclusion: Was the goal of the experiment achieved? Was the hypothesis (model) substantiated or disproved? How well did the experimental design work toward achieving the goal? What should have been done differently? What should be done next?

Some researchers like to make their conclusions especially clear by numbering them. The more important ones are placed first, and the less important conclusions are last. This idea is a good one. When you sit down to write a technical paper or report that is based on your lab notes, having the conclusions clearly laid out in front of you is indeed a big help. Often you can write the abstract for your paper by merely copying the conclusion section of your notebook entry. Looking at it another way, your conclusion section should contain all the information that you would put into an abstract describing the work.

Recording Ideas

One of the most important uses of the notebook for the industrial scientist is to record original ideas. If the idea leads to a patentable invention, you will need to demonstrate when the idea occurred, who was the inventor, who contributed to the development of the idea, and how diligent was the inventor in reducing the idea to practice. These particular aspects of notekeeping are dealt with in detail in Chapter 7. For now, you should realize that you must record any novel idea that you have as soon as the idea occurs. Indicate the inspiration for the idea. State whether you discussed the idea with other people and indicate what, if anything, was their contribution.

If you do write an idea onto a scrap of paper, be sure to either paste or tape that note into your notebook as soon as possible. Sign and date the note and have your signature witnessed. Alternatively, write the idea into the notebook as soon as you can.

Write the description of your idea as plainly as possible. Often an idea will be crystal clear in your head but very difficult to describe to others. A useful approach is to write in especially short, declarative sentences after preparing a short outline of how your invention or new method will work. Drawings can be especially helpful even if they are simple. When you ask someone to read and witness your notes, also ask them if the explanation is clear and complete or if additional description should be added.

Literature Surveys

Each scientist has a favorite way of making notes while preparing a literature review. If the review is brief and intended to support a specific experiment, it should be included in the introductory section of the notebook writeup. If the literature review will be lengthy, such as a stand-alone publication, it may be better to use notecards, file folders, or a loose-leaf binder to hold the information. These methods permit easy rearrangement of the material.

If you are starting an especially detailed and lengthy literature survey, you can set up a notebook with index tabs to mark groups of blank pages that will be filled in during the course of your research. Each section will be devoted to a different topic of the literature review. This method keeps related notes together and minimizes the chances of losing a piece of paper with important notes. You can easily carry your notebook into the library or wherever your research takes you. On the negative side, this method has two significant disadvantages. First, using a bound notebook inhibits you from physically rearranging the notes into a more logical order when preparing the actual draft of the report. Second, you will be jumping to different sections of the notebook and your notes will not be in chronological order. This point may be unimportant for the academic researcher but may cause some trouble for the industrial researcher who is concerned about demonstrating continuity and diligence in the notebook.

Computer programs are available to record notes and to create outlines and indexes for reports. Word processing programs are commonly used to write the entire text of scientific papers. If you have access to one of these programs, use it in preference to a notebook for preparing lengthy literature review articles.

The Notebook as a Training Record

One of the handiest uses of a notebook is to keep track of new methods, techniques, procedures, or other lessons you learn on the job. For example, when your boss says that you will be trained to operate a new instrument, bring along a notebook to record the instructions that you will receive. Even though you may be given an instruction manual of information sheets about the apparatus, you will also gain insights into the machine's operation from the instructor, some of which are never formally written down. These little "hints" are often just as important (if not more so!) than the manufacturer's original instructions. Ask questions and take the time to write down the answers. The act of writing in the notebook helps you to remember. Then, bring the notebook with you when you begin to operate the equipment and continue to record useful information about the machine's quirks. I have notebooks in which I continue to make such notes several years after having been introduced to a new instrument.

Another reason to record your lessons in a notebook is that you can have the instructor look over the notes to be sure that your understanding of the instructions is correct. If you then follow your written operating procedures, you will be less likely to get yourself into trouble with the boss.

Notebooks with More Than One Author

In some situations, more than one person will make entries in the same notebook. Experiments that run continuously, such as pilot plant or scale-up operations, often have workers on hand around the clock to monitor the equipment. The job of taking periodic readings of instrument dials has largely been taken over by automatic recording of the information. Still, a person is often required to make notes on the performance of the equipment and the appearance of the product.

If several people will be making notes in the same book, get together before beginning the project and briefly discuss any special aspects of the notekeeping process such as data tables, symbols, or shorthand notations to be used. Few things are more frustrating than beginning a shift and not being able to decipher the previous person's notes. Because you are writing especially for others, write legibly. If in doubt, spell out your comments rather than use abbreviations. Note the time of an event by using a 24-hour clock (military time) or by including a.m. or p.m.

Each person who makes a notebook entry should be sure to initial the entry so that any subsequent questions about the work will be addressed to the right person.

As mentioned in Chapter 4, some companies permit only professional staff to make notebook entries if the work may lead to a patentable invention. Technicians are permitted to enter data if they are under the direct supervision of a professional. The entry is immediately signed and dated by the technician, but the bottom of the notebook page is signed, dated, and witnessed by professional staff members. If technicians have not been thoroughly instructed in the legal ramifications of notekeeping, give them specific instructions just before they are to record their work. Probably the most important instruction is for the technician to avoid any sort of speculation and to record the facts exactly as they are observed.

Many Projects in One Notebook

If you are engaged in technical service work, technical marketing support, materials failure analysis, or forensic science, you will find yourself involved in many brief, remarkably different projects, each lasting from a few hours to a few days. What is the best way to set up the notebook for this sort of work?

Begin with a preface (at the front of the notebook) that describes your position and the type of work that will be recorded in the notebook. List the instruments, reagents, and procedures you will use. A table of abbreviations for these common items will shorten your daily notekeeping. If other people or departments routinely get involved in your work, then include their names and job titles in your table of abbreviations.

Start recording each project on a new page. Begin as you would for a research project by writing the date, project number, and project title at the top of the page. If you know the name of the client or department for whom the work is being performed, make a note of that, too. When you start the work, make an entry for this project in the table of contents to help you find the proper page quickly when you need to record additional information.

Next, for the purpose of an introduction, write a few sentences stating the specific problem that you have been asked to solve. Do not speculate on the cause of the problem until you have sufficient data to make an educated guess. Idle speculation at this point can haunt you later if, for example, you are called to testify about the work.

Many technical service projects call for careful documentation of samples received for examination. Your organization may have a quality assurance program that spells out the specific routine to follow in receiving and identifying samples before work can begin. If documentary photographs or other information has been supplied to you, be sure that you record that fact in your notebook and cross-reference the files in which that information is stored.

Record your work on the problem by following the guidelines presented in the previous sections. Have your data and calculations checked by a coworker if needed. You must remember to note instrument calibrations and procedure validations if your results might be used in litigation. Describe any communication that you have with the client or with a colleague that bears on solving the problem. Clearly label the sections in your notebook so that no confusion exists as to which section contains factual data and which section contains your opinions and intepretation of these data.

Finally, make a clear and carefully worded summary of your work under the heading "Results" or "Conclusion." Preface your conclusion by a phrase such as "The evidence suggests..." or "I found evidence of...." Draw together a summary of your observations that led you to the conclusion. Do not fix blame for the problem on any person or organization unless you have additional, substantiated information to do so. Stick to a discussion of your specific evaluation of the materials in question. Sign and date after the last line of your entry and have your notes witnessed by the appropriate person. If samples or related documents are to be stored, make a note at the end of your notebook entry of where and under what conditions the materials are stored.

To make your job a little easier in the future, you may want to build a keyword cardfile or computer-based index with references to the problems you have studied and their resolution. Annotating the table of contents or writing an index at the back of the notebook can serve the same purpose.

A Final Word on Grammar, Style, and Tone

The purpose of writing notes is to someday read them. In the words of Reginald Kapp: "In the end the proud scientist or philosopher who cannot be bothered to make his thought accessible has no choice but to retire to the heights in which dwell the Great Misunderstood and the Great Ignored, there to rail in Olympic superiority at the folly of mankind." To protect their work, some notewriters are especially cryptic. Leonardo da Vinci wrote his notes in mirror style so that they appeared to be gibberish to the average person. The fictional inventor, Tom Swift, Jr., wrote his notes in a code that only he and his scientist-father could understand. However, the average student and working scientist should write plainly and clearly. (Protect the physical contents of the notebook in accordance with the principles given in Chapter 4.)

You will find no finer and more succinct tutor in proper writing than *The Elements of Style* by William Strunk and E. B. White. Read it.

Examples of Notebook Entries

T his chapter provides examples of the proper method of entering a variety
of laboratory notes. Figure 6.1 is an example of an apparatus sketch.
Dimensions are given, and the direction of gas flow is shown. A detailed
explanation of how to use the apparatus would be found in the text of the
experiment. Items shown in the figure can also be numbered for reference in the
text.

Figure 6.2 contains seven pages from an industrial researcher's notebook
describing an organic synthesis that may lead to patentable new products. In a real
notebook, such notes are often scattered among pages describing other work.

Figure 6.3 contains four notebook pages that show data tabulation, plotting
of results, and calculations. *Note* the headings at the beginning of each section.
The reference to a previous procedure eliminates a redundant description. A
discussion of the results normally would follow the calculations.

The two pages in Figure 6.4 illustrate the process of "thinking in the
notebook." They show the very beginning of a research project. Supplementary
information will be collected in other files, and computers will be used to write
the actual proposal and to calculate budgets. However, the notebook is the best
place to record questions, answers, and ideas about a project.

A format for recording quality-control data is provided in Figure 6.5. *In
advance,* set up as many pages as required in the same format. A rubber stamp can
be made with the appropriate headings so that only the blanks need to be filled in.
If each page is set up the same way, the results can be easily read and transcribed
for analysis. Alternatively, the form can be printed or photocopied onto separate
sheets of paper called *bench sheets.* Using the notebook has the obvious advantage
of keeping the pages bound together, in sequence.

Students should record observations of a teacher's lecture-demonstration in a
notebook (Figure 6.6). This exercise is good practice for descriptive writing. *Note*
the use of personal pronouns to tell the story.

0906–5/85/0081/$06.25/1
©1985 American Chemical Society

Figure 6.7 is an example of a page from a personal diary. The subjective viewpoint of the entries would be out of place in a proper laboratory notebook.

Even if instruments are computer controlled, a hand-written instrument logbook is easy to keep, always available, and portable. The instrument's recent performance and calibration can quickly be found. An example of a page from an instrument logbook is shown in Figure 6.8.

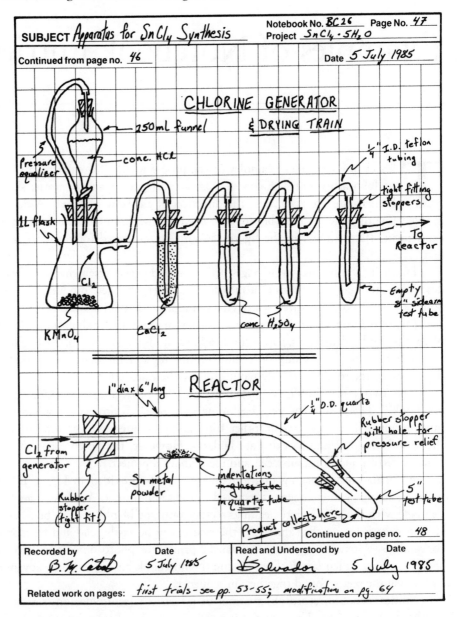

Figure 6.1

| SUBJECT BUG REPELLENTS | Notebook No. *CP1144* Page No. *67* |
| | Project *CP-042-85* |

Continued from page no. — Date May 2, 1985

INTRODUCTION - SYNTHESIS OF POTENTIAL REPELLENTS

This is the start of a new project to synthesize & test potential new insect repellents. Dr. Carlson has asked me to do a literature search on the effectiveness of anthranilic acid esters; much of this information is probably proprietary—we will have to "re-invent the wheel." My plan is to ① do a computer-based literature search from my own terminal using DIALOG, and then trace back any references prior to 1967 by hand; ② Synthesize a few esters and give them to Carol Wilson's group for testing their effectiveness. She will only need ~100g of each compound.

Since this project may eventually lead to a patented product, I must take pains to see that all reagents are characterized before use, that the products are also carefully characterized, and that my notes are properly witnessed.

May 7, 1985

Earlier today, I worked on the literature review. I found several references to the | Continued on page no. 68

| Recorded by | Date | Read and Understood by | Date |
| CoDisley | 7 May 1985 | Saluador | 14 May 1985 |

Related work on pages: _____

Figure 6.2 Page 1 of 7.

| SUBJECT Bug Repellents - Background | Notebook No. CP-1144 Page No. 68 |
| | Project CP-042-85 |

Continued from page no. 67   Date 7 May 1985

(Literature review, cont'd)

synthesis of anthranilic acid esters, but only the propyl ester
was reported to be a repellent (against Haltica chalybea).
Copies of relevant abstracts and papers will be filed in
the project CP-042-85 folders, kept in room A618.

10 May 1985

No additional work on this project this week -- I am
waiting for the literature to arrive from the data-base
suppliers.

14 May 1985

Received and read most of the relevant literature.
I will begin synthesis of the methyl and propyl esters.
EXPERIMENTAL PLAN

Conventional Fischer esterification with alcohol and
HCl has been reported to produce only 50-70% yields of
methyl anthranilate. Brenner & Huber (Helv. Chim. Act. 36, 1112,
1953) used MeOH and thionyl chloride to prepare methyl
esters of amino acids of the general form $R-C\begin{smallmatrix}COOH\\NH_2\end{smallmatrix}$.
I will follow their procedure to prepare the
methyl- and n-propyl esters of anthranilic acid.

Continued on page no. 69

| Recorded by   Date | Read and Understood by   Date |
| G. Drinny   14 May '85 | Salvador   14 May 1985 |

Related work on pages: _____

Figure 6.2   Page 2 of 7.

SUBJECT Bug Repellents - Expt'l Plan

Notebook No. CP-1144 Page No. 69
Project CP-042-85

Continued from page no. ~~68~~ 68   Date 14 May 1985

The outline of their procedure is:

1.) Cool MeOH (or $CH_3CH_2CH_2OH$) to ~ -10°C and add $SOCl_2$.
2.) Add anthranilic acid; slowly warm & reflux
3.) Evaporate solvent and recrystallize from ethanol/ether.
4.) This HCl salt will then be dissolved in water,
    neutralized with $NaHCO_3$, & extracted with ether.

Table of Properties

	substance	F.W.	FORMULA	MP	Ref.
1.	anthranilic acid	137.13	$C_7H_7NO_2$	~144°c	Sadtler #2703
2.	" phenylurea deriv.	—	—	181°c	Shriner, etal pg. 246
3.	" acetyl deriv.	—	—	185°c	"
4.	" methyl ester	151.16	$C_8H_9NO_2$	23-24°c	Sadtler #897
5.	" propyl ester·HCl	215.68	$C_{10}H_{13}NO_2 \cdot HCl$	~160°c	Arch. Pharm 263, 481
6.	" propyl ester	179.22	$C_{10}H_{13}NO_2$	~7°c	" (1925)

Theoretical Composition

	C(%)	H%	N(%)	O(%)
1.	61.3	5.1	10.2	23.3
4.	63.6	6.0	9.3	21.2
6.	67.0	7.3	7.8	17.9

Continued on page no. 70

Recorded by	Date	Read and Understood by	Date
G. Denny	14 May '85	Saluador	14 May 1985

Related work on pages: _____

Figure 6.2   Page 3 of 7.

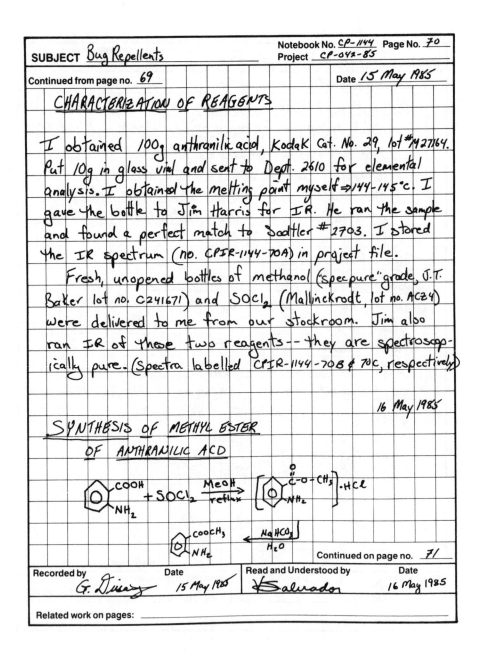

| SUBJECT Bug Repellents | Notebook No. CP-1144 Page No. 70 |
| | Project CP-042-85 |

Continued from page no. 69   Date 15 May 1985

CHARACTERIZATION OF REAGENTS

I obtained 100g anthranilic acid, Kodak Cat. No. 29, lot #M27164. Put 10g in glass vial and sent to Dept. 2610 for elemental analysis. I obtained the melting point myself ⇒144-145°C. I gave the bottle to Jim Harris for IR. He ran the sample and found a perfect match to Sadtler #2703. I stored the IR spectrum (no. CPIR-1144-70A) in project file.

Fresh, unopened bottles of methanol ("spec pure" grade, J.T. Baker lot no. C241671) and SOCl₂ (Mallinckrodt, lot no. AC24) were delivered to me from our stockroom. Jim also ran IR of these two reagents -- they are spectroscopically pure. (Spectra labelled CPIR-1144-70B & 70C, respectively)

16 May 1985

SYNTHESIS OF METHYL ESTER
OF ANTHRANILIC ACID

Continued on page no. 71

| Recorded by | Date | Read and Understood by | Date |
| G. Di759 | 15 May 1985 | Salvador | 16 May 1985 |

Related work on pages: _____

Figure 6.2   Page 4 of 7.

SUBJECT Bug Repellents

Notebook No. _CP-1144_ Page No. _71_
Project _CP-041-85_

Continued from page no. _70_

Date _16 May 1985_

$\left(\text{Synthesis of } \underset{\underset{NH_2}{}}{\overset{COOCH_3}{\bigcirc}} \text{, continued}\right)$

Trial quantities needed (based on Brenner & Huber's work)

6.0 mL (0.15 moles) methanol
1.15 mL (0.016 moles) SOCl$_2$
1.95 g (0.015 moles) anthranilic acid

Experimental Procedure & Observations

I set up a 20 mL roundbottom flask over ice/KCl bath with a magnetic stirrer. Added 6.0 mL methanol and cooled until thermometer indicated -7°C. Added 1.15 mL SOCl$_2$ dropwise with stirring ($\frac{1}{2}$" teflon bar), keeping the temp below +5°C. Next I added 1.95 g anthranilic acid in ~0.2 g portions with stirring. A white paste formed which stopped the stirrer. I removed the ice/KCl bath, replaced it with a water bath & heater. Began heating slowly. After ~20 minutes, temperature was ~70°C and the paste began to dissolve, the solvent began to boil. Put a condensor and CaCl$_2$ drying tube on the flask and refluxed 2 hrs.

Continued on page no. _72_

Recorded by	Date	Read and Understood by	Date
G. Drury	16 May 1985	Salvador	16 May 1985

Related work on pages: _____

Figure 6.2 Page 5 of 7.

| SUBJECT Bug Repellents | Notebook No. CP-1144 Page No. 72 |
| | Project CP-045-85 |

Continued from page no. 71   Date 16 May 1985

(synthesis of methyl anthranilate, cont'd)

The temperature of the refluxing solution was $70 \pm 3°C$ for the whole time. After 2 hrs, the solution was pale yellow, with some suspended particles. I filtered the hot solution through a fine glass frit and "roto-vapped" the solvent. After ½ hr, the flask contained a grey residue.

This residue was taken up in ~50 mL anhydrous ethanol (Baker, lot no. C241998) and filtered. I cooled the filtrate in an ice bath and added ~50mL diethyl ether (Kodak, lot no. X247). White crystals precipitated which I then filtered and sucked dry. This is the methyl ester hydro-chloride salt of anthranilic acid. Labelled "CP-1144-72A." Yield = 1.62 g. Stored in glass vial at room temp.

17 May 1985

(This morning I received analysis of the sample of anthranilic acid sent to analyt. lab on 15 May (this book, pg 70): C = 61%, H = 5.1%, N = 10.1%, O = 23.1%. Excellent agreement with expected composition.)

To make the methyl ester from the salt, I put 1.0g of CP-1144-72A in 15 mL deionized ($18 M\Omega$-cm) water and stirred to dissolve. Added $NaHCO_3$ until basic to litmus.

Continued on page no. 73

| Recorded by | Date | Read and Understood by | Date |
| G. Disöng | 17 May 1985 | Salvador | 17 May 1985 |

Related work on pages: _____

Figure 6.2 Page 6 of 7.

| SUBJECT Bug Repellent | Notebook No. _CP-1144_ Page No. _73_ |
| | Project _CP-045-85_ |

Continued from page no. _72_   Date _17 May 1985_

(synth. of methyl anthranilate, cont'd)

Gas bubbles (must be CO_2) evolved gently. Extracted the solution 2 x 30 mL diethyl ether. I combined the extracts and evaporated the ether using gentle steam heat. The product was a light-yellow oil which should be the methyl ester of anthranilic acid. Yield ≅ 2 mL, labelled "CP-1144-73A." Sent out to org. analyt. lab for IR, NMR, and analysis. If the material is sufficiently pure, I will scale up and make a couple of hundred grams for Dr. Wilson's tests.

Summary of Work-To-Date

Thus far, I have synthesized a small quantity of methyl anthranilate. The method seems to work o.k.; If the product is sufficiently pure, I will make enough for the bug repellent tests. Then, I will proceed to make some other esters by the same method.

Continued on page no. _____

| Recorded by G. Dewey | Date 17 May 1985 | Read and Understood by Salvador | Date 17 May 1985 |

Related work on pages: _____

Figure 6.2 Page 7 of 7.

| SUBJECT *FREEZING PT. DEPRESSION* | Notebook No. *1123* Page No. *24* |
| | Project *Detm'n of Avogadro's No.* |

Continued from page no. *23* | Date *16 March 1985*

OBSERVATIONS & DATA

I set up the apparatus described in Shoemaker, et al, "Experiments in Physical Chemistry", p. 180 (1974). The high-precision Beckman thermometer is serial no. L2634, readable ±0.01°C. The procedure was identical to last week's (see this notebook, pp. 16-19.) and I collected the following data:

TRIAL	# grams solute added
1	-0- "pure" H_2O (18 MΩ-cm)
2	0.3279
3	0.6626
4	1.0008

Time (min.)	TEMP (°C)			
	Trial 1	Trial 2	Trial 3	Trial 4
0.5	0.05	-0.28	-0.23	-0.24
1.0	-0.05	-0.40	-0.40	-0.44
1.5	-0.15	-0.46	-0.57	-0.64
2.0	-0.22	-0.51	-0.63	-0.83

Continued on page no. 25

Recorded by	Date	Read and Understood by	Date
H.D.Kan	16 March '85		

Related work on pages: _____

Figure 6.3 Page 1 of 4.

| SUBJECT *Freezing Point Depression* | Notebook No. *1123* Page No. *25* |
| Project *Avogadro's No.* |

Continued from page no. *24*   Date *16 March 1985*

(Data collection, continued)

Time (min.)	Temp (°c)			
	Trial 1	Trial 2	Trial 3	Trial 4
2.5	−0.28	−0.55	−0.85	−1.01
3.0	−0.35	−0.60	−1.02	−1.18
3.5	−0.44	−0.64	−1.19	−1.32
4.0	−0.49	−0.64	−1.32	−1.50
4.5	−0.50	−0.66	−1.45	−1.70
5.0	−0.60	−0.70	−1.60	−1.84
5.5	−0.18	−0.70	−1.75	−2.00
6.0	−0.13	−0.71	−1.85	−2.18
6.5	−0.11	−0.70	−1.95	−2.30
7.0	−0.11	−0.71	−2.10	−2.48
7.5	−0.11	−0.71	−2.23	−2.60
8.0	−0.11	−0.71	−2.34	−2.75
8.5	No more readings	No more readings	−2.45	−2.80
9.0			−1.80	−2.02
9.5			−1.34	−1.93
10.0			−1.32	−1.91
10.5			−1.32	−1.91
11.0			−1.32	−1.91

Continued on page no. *26*

Recorded by *J. M. Kaw*   Date *16 March '85*   Read and Understood by   Date

Related work on pages: _____

Figure 6.3   Page 2 of 4.

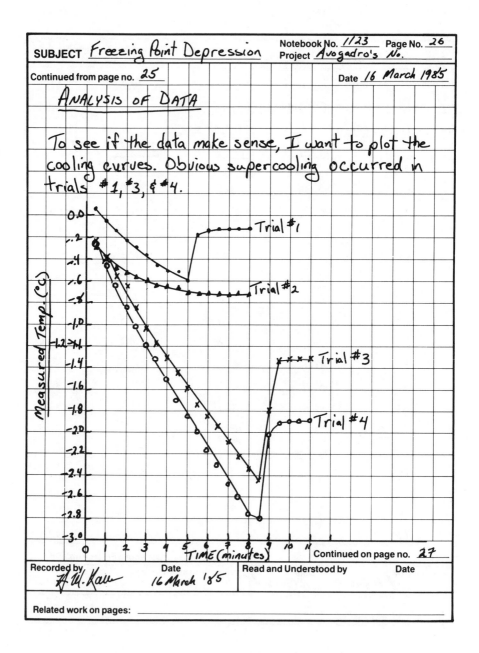

Figure 6.3   Page 3 of 4.

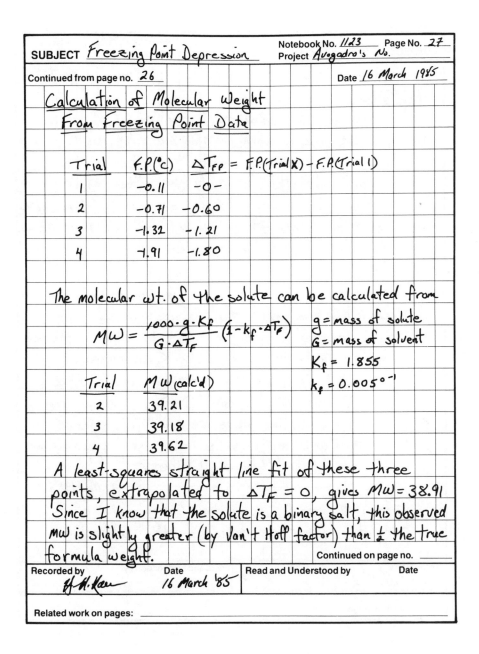

Figure 6.3 Page 4 of 4.

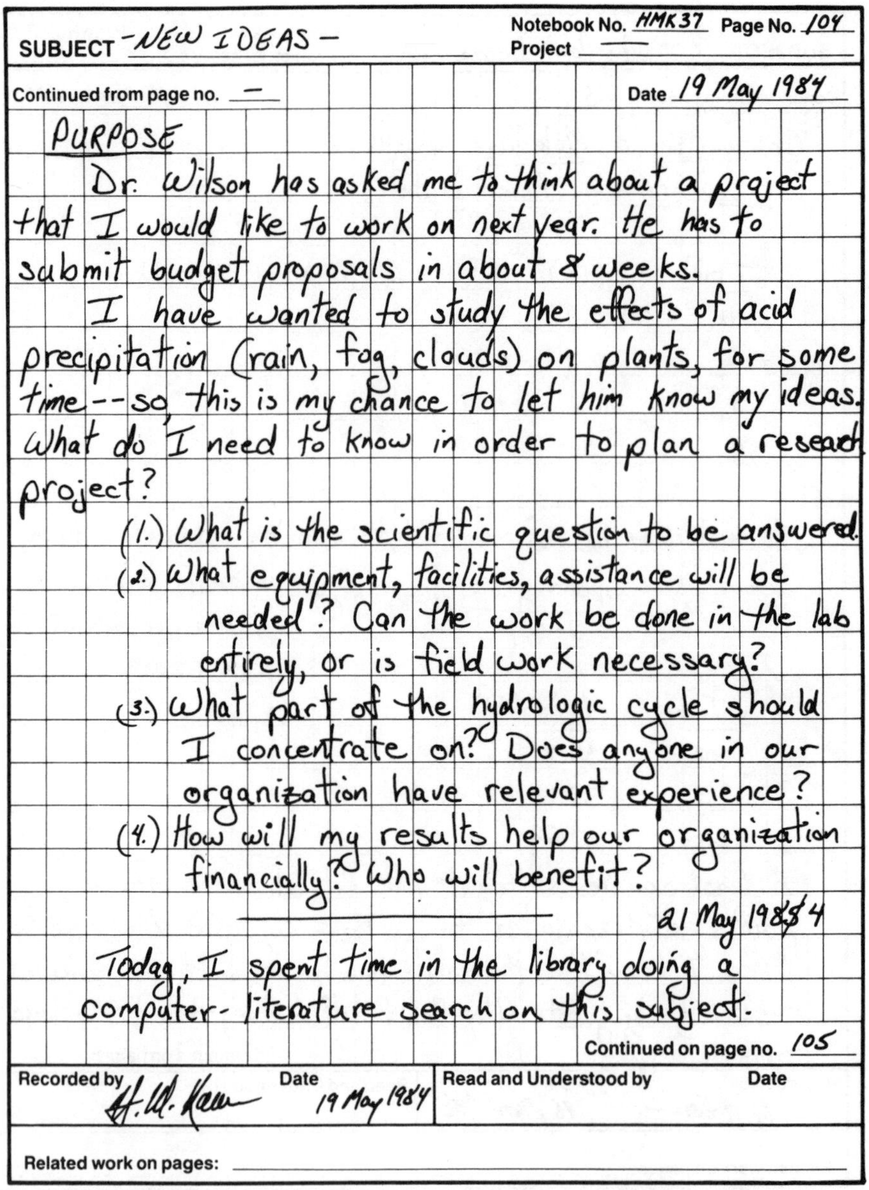

| SUBJECT —NEW IDEAS— | Notebook No. HMK37 Page No. 104 |
| | Project _____ |

Continued from page no. —   Date 19 May 1984

PURPOSE

Dr. Wilson has asked me to think about a project that I would like to work on next year. He has to submit budget proposals in about 8 weeks.

I have wanted to study the effects of acid precipitation (rain, fog, clouds) on plants, for some time -- so this is my chance to let him know my ideas. What do I need to know in order to plan a research project?

(1.) What is the scientific question to be answered.

(2.) What equipment, facilities, assistance will be needed? Can the work be done in the lab entirely, or is field work necessary?

(3.) What part of the hydrologic cycle should I concentrate on? Does anyone in our organization have relevant experience?

(4.) How will my results help our organization financially? Who will benefit?

21 May 1984

Today, I spent time in the library doing a computer-literature search on this subject.

Continued on page no. 105

| Recorded by H.W. Kaw Date 19 May 1984 | Read and Understood by   Date |

Related work on pages: _____

Figure 6.4   Page 1 of 2.

SUBJECT *ACID RAIN PROPOSAL*

Notebook No. *HMK 37*   Page No. *105*

Project _____

Continued from page no. *104*   Date *21 May 1984*

    The abstracts of relevant articles are being sent to me by mail and I will begin a file of them. I have narrowed my thoughts to the study of the effects of acid cloud vapor, and prevention of damage to plants at moderately high elevations.

BACKGROUND

    Much work has apparently been done on the treatment of acidified lakes to restore proper pH, especially in Sweden & Germany. It is costly, but can be done because of the relatively small surface area of lakes compared to the forests. However the forests that exist above ~1000 m in the Northeastern U.S. are especially vulnerable to acid attack — the soils are thin, cannot neutralize the acid rain & fog, and the coniferous trees have branches and leaves that catch cloud moisture very effectively.

    A possible topic, therefore, is this: Are any species of conifer that grow in sub-alpine environments less susceptible to acid precipitation than other species? If so, why?!

Continued on page no. *106*

Recorded by *H. M. Kann*   Date *21 May 1984*

Read and Understood by   Date

Related work on pages: _____

Figure 6.4 Page 2 of 2.

| SUBJECT QUALITY CONTROL RECORD | Notebook No. B-124   Page No. 44 |
| | Project Dept. 021-QC |

Continued from page no. —   Date 2 May 1985

IDENTIFICATION – ION CHROMATOGRAPHY RESINS

MATERIAL:

BATCH No.:

DATE PRODUCED:

ANALYSIS

BEAD SIZE DISTRIBUTION

    Percent + No. 100 U.S. Std. Sieve:

    "   + No. 200 "   "   "

    "   + No. 325 "   "   "

    "   – No. 325 "   "   "

COLOR OF WATER EXTRACT:

pH of water slurry:

PERFORMANCE

TEST FIXTURE No.:

RETENTION TIME FOR Cl^-:

RETENTION TIME FOR SO_4^{2-}:

COMMENTS:

PASSES TESTS? _____   Continued on page no. _____

| Recorded by | Date | Read and Understood by | Date |

Related work on pages: _____

Figure 6.5

| SUBJECT CHEM 102 - Demo | Notebook No. V.S. Page No. 6 |
| Project |

Continued from page no. ——   Date 14 Feb. 1985

Dr. Norris's Lecture Demo —
    Reaction of Sodium with Water

Dr. Norris began by filling a petri dish with
deionized water. Then she added 3 drops of
~~fenoth~~ phenolphthalein indicator -- the water
stayed colorless. Next, she showed us a glass
jar filled with pieces of sodium in a colorless
liquid. The liquid looked and smelled like kerosene.
    Next, Dr. Norris took out a piece of the sodium.
It was mostly covered with a whitish-grey coating.
She patted it dry with a paper towel and then
cut a pea-size lump with a knife. It was
soft! The inside (cut surface) looked metallic but
not real shiny. She put the dish with the water
on the overhead projector and turned it on so we
could all see through the water, projected onto
the overhead screen.
    When she put the piece of sodium into the water,

Continued on page no. 7

| Recorded by   Date | Read and Understood by   Date |
| Salvador   14 Feb. 1985 | —— |

Related work on pages: _____

Figure 6.6 Page 1 of 3.

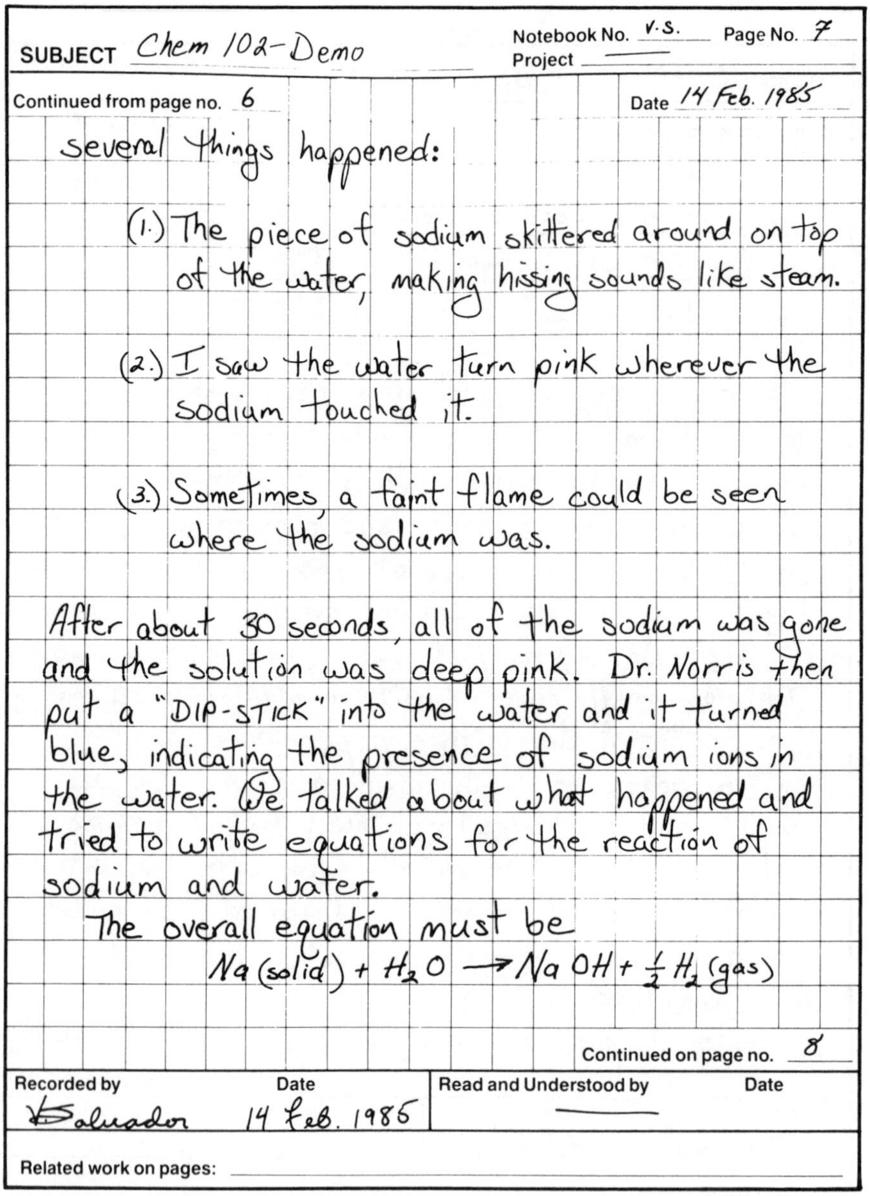

SUBJECT _Chem 102-Demo_   Notebook No. _v.s._   Page No. _7_
                                 Project _____

Continued from page no. _6_   Date _14 Feb. 1985_

several things happened:

(1.) The piece of sodium skittered around on top
      of the water, making hissing sounds like steam.

(2.) I saw the water turn pink wherever the
      sodium touched it.

(3.) Sometimes, a faint flame could be seen
      where the sodium was.

After about 30 seconds, all of the sodium was gone
and the solution was deep pink. Dr. Norris then
put a "DIP-STICK" into the water and it turned
blue, indicating the presence of sodium ions in
the water. We talked about what happened and
tried to write equations for the reaction of
sodium and water.
   The overall equation must be
$$Na \,(solid) + H_2O \longrightarrow NaOH + \tfrac{1}{2}H_2 \,(gas)$$

Continued on page no. _8_

Recorded by   Date   Read and Understood by   Date
Salvador   14 Feb. 1985   _____

Related work on pages: _____

Figure 6.6   Page 2 of 3.

| SUBJECT _Chem 102-Demo_ | Notebook No. _V.S._ Page No. _8_ Project _____ |

Continued from page no. _7_   Date _14 Feb. 1985_

<u>Conclusions</u>

   The sodium dissolved to give sodium ions; it was oxidized from Na to Na^+. The hydrogen was reduced from $2H^+$ to H_2 gas which burned in air. So, the OH^- combined with Na^+ to stay in the solution (dissolved), making the water very alkaline & turning the indicator to pink.

   The sodium had been stored in Kerosene without dissolving or burning up. So, the sodium is strong enough to reduce hydrogen ions in water to hydrogen gas, but the sodium doesn't react with the hydrogens that are part of the Kerosene molecules. I have to think about that one -- why not?!

   Finally, I thought about why the sodium stayed on top of the water. The density of Na is $0.97 g/cm^3$, which is less than water, so it floats!

Continued on page no. _____

| Recorded by | Date | Read and Understood by | Date |
| Saluador | 14 Feb. 1985 | _____ | |

Related work on pages: _____

Figure 6.6 Page 3 of 3.

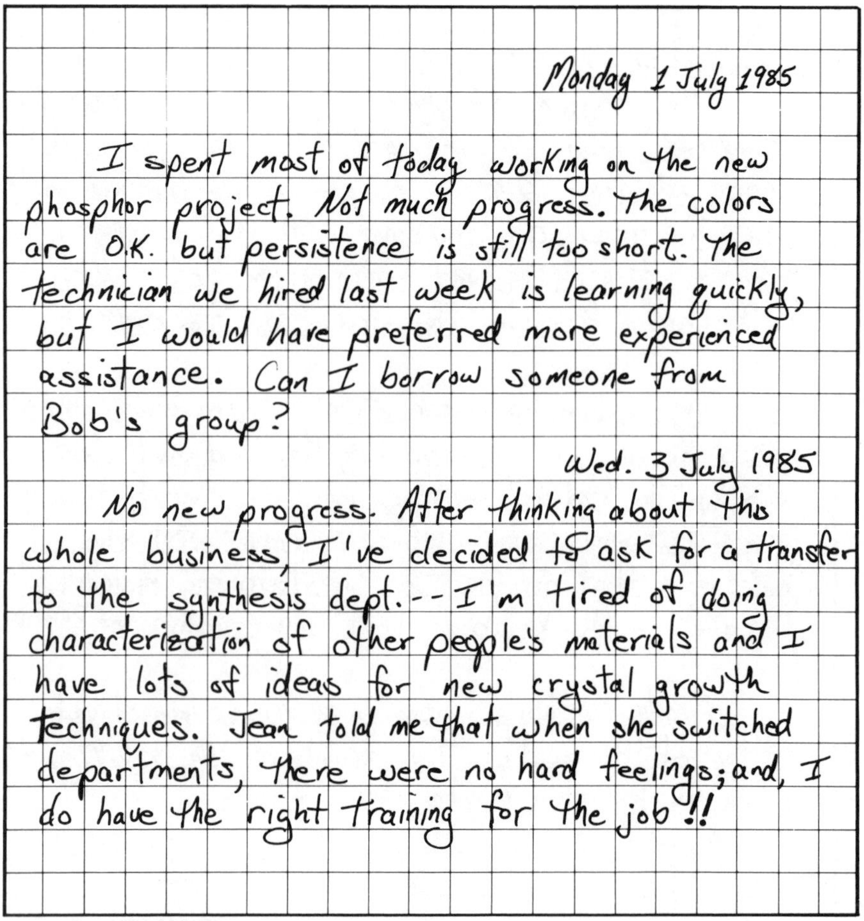

> Monday 1 July 1985
>
> I spent most of today working on the new phosphor project. Not much progress. The colors are O.K. but persistence is still too short. The technician we hired last week is learning quickly, but I would have preferred more experienced assistance. Can I borrow someone from Bob's group?
>
> Wed. 3 July 1985
>
> No new progress. After thinking about this whole business, I've decided to ask for a transfer to the synthesis dept. --- I'm tired of doing characterization of other people's materials and I have lots of ideas for new crystal growth techniques. Jean told me that when she switched departments, there were no hard feelings; and, I do have the right training for the job!!

Figure 6.7

SUBJECT FTIR LOGBOOK			Notebook No. 21466 Page No. 12
Continued from page no. 11			Date _____

DATE	PROJECT	OPERATOR	COMMENTS
28 June '85	XL-1493-C	*H.M. Kaw*	2 pvc pipe fragments; working O.K.
"	New coating	*L.S. Liner*	O.K.
"	―	*R. Dose*	Calibration check
29 June '85	A291-C4	*J.D. Scott*	computer crash-power failure; Restarted o.k.
1 July 1985	XL-1494-C	*H.M. Kaw*	coating on mylar by ATR; cleaned the mirrors
"	"	*H.M. Kaw*	2 poly. substrates for Dr. Wilson; O.K.
"	Dyes	*Theo. Kretos*	O.K.
"	Q.A.	*George N'Dan*	O.K.
"	Q.A.	*Tom Sharf*	O.K.
2 July '85	―	calibration by	manufacturer - see Q.A. file -
"	Dyes	*Theo. Kretos*	O.K.
"	Hyd. Kinet.	*S. Jones*	Time study - 2 solution's; Runs fine
3 July '85	A294-C4	*J.D. Scott*	"T" knob fell off; will order new one
"	―	*G. Berther*	O.K.

Figure 6.8

Patents and Invention Protection

Every researcher dreams of making a breakthrough discovery or invention that will establish a place in history and ensure financial rewards. Occasionally that event happens. More often than not, however, the day-to-day grind of laboratory investigation eventually results in an improvement in an existing process or substance. In either situation, the inventor (or, more likely, the inventor's employer) must choose how to profit from the invention and how to protect the company's investment in the invention. Two options are usually available: trade-secret protection or patent protection. In the first option, the invention may be kept secret and used by the inventor's company without disclosing to the public how the invention is made. This approach is most likely to succeed if the invention is extremely difficult to duplicate by reverse engineering. (The formulation for Coca-Cola is probably the most famous example of a well-kept trade secret.) However, if someone figures out how to make the invention, the value of the secret will be lost.

A patent, on the other hand, gives its owner the right to exclude others from making, using, or selling the invention for the length of time that the patent is in force. This right is given by the federal government in exchange for the inventor's public disclosure of how to make and use the invention. A patent can be issued only to the true inventor, and the invention must be new, unobvious, and useful. A trade secret may be profitable for as many years as the secret can be kept, whereas the life of a patent is relatively short and limited by statute.

The notebook serves three major functions in the area of securing and protecting patents. First, the notebook provides some of the information needed to decide whether to apply for a patent. Second, the notebook is one of the major sources of information needed to prepare the patent application. Third, if the

0906–5/85/0103/$06.00/1

patent is challenged by someone else claiming to be the inventor, the notebook will be used as part of the original inventor's evidence of when the invention was conceived and when it was first demonstrated.

You will never know in advance when your notes might be needed to prove that you were the original inventor and that you conceived or built an invention on some particular date. Keep notebooks according to sound principles at all times, not just when your work is likely to "get patents." In this chapter, we will examine how the notebook is used to help secure patents and invention protection. The general principles presented are applicable in most industrial organizations. However, you should be aware of (and follow) specific procedures recommended by your company's patent counsel.

Patent or Trade Secret Protection?

The research scientist must realize that the decision to practice an invention as a proprietary product or to patent the invention is a business decision. The scientist can best assist the managers and lawyers who make that decision by supplying complete details of the invention's conception and subsequent reduction to practice. Those two terms require some explanation. *Conception* is the formulation of the invention in the mind of the inventor. (If several people contribute ideas that lead to the concept of the invention, they may be coinventors.) *Reduction to practice* is the actual carrying out of the invention, such as the first experimental demonstration proving that the invention works. Building a working version of the invention simply to show that it does work may be impractical. If so, the filing of the patent application is termed *constructive reduction to practice*.

Corporate management must assess the economic feasibility of an invention to determine whether or not to file a patent application. Maynard (1) lists ten questions to consider in deciding whether or not to file a patent application. Five of these ten questions should be answerable simply by perusing the inventor's notebook: (1) What is the invention? (2) When was the invention made and where is the experiment recorded? (3) Has the invention been disclosed publicly? (4) What is known about prior work in the field? (5) How complete are the data? Responsibility rests on the inventor's shoulders to make sure that research records are complete, timely, properly witnessed, and in accordance with the employer's guidelines.

If the decision is made to go ahead with a patent application, the inventor (or someone acting on the inventor's behalf) prepares a memorandum for the patent attorney called an *invention disclosure*. An outline that can be used in writing such a disclosure is given in Appendix II of Reference 2. In preparing an invention disclosure, the following information is usually needed: the names and addresses of the inventors, the date of conception of the invention, the date when a working model of the invention was demonstrated, a description of the invention, an explanation of why the invention is novel, the uses of the invention, and a list of how and when the invention has been discussed or shown to others. Most of this information comes straight from the inventor's notebooks. All information in the invention disclosure should be backed up by primary documents in project files, such as the inventor's original notes and data.

The attorney will prepare a draft of specifications and claims to be used in the formal patent application that is based on the invention disclosure and consultations with the inventor. When reviewing these draft documents, the inventor must check for technical accuracy and to be sure that information in the documents is consistent with research records including lab notebooks and project files.

Important Notekeeping Details

United States patent law gives priority to the first person to conceive of an invention and then proceed to make a working version of the invention. According to Title 35 of the United States Code, Section 102(g):

> In determining priority of invention there shall be considered not only the respective dates of conception and reduction to practice of the invention, but also the reasonable diligence of one who was first to conceive and last to reduce to practice, from a time prior to conception by the other.

If another party files a patent application for substantially the same invention, the Patent Office will begin an *interference proceeding* that will result in a ruling as to whom the patent should be awarded. (Interference proceedings can be started in several ways as can later court challenges to the patent.) Only about 1% of the patent applications filed with the Patent Office become involved in an interference proceeding (3). That number may seem like a small percentage, but over 100 thousand patent applications are received every year. Thus, over 1000 interference proceedings are instituted annually. To provide the most useful evidence for your case, pay special attention to the following notekeeping details: (1) recording conception of an invention, (2) continuity of the record, (3) witnessing, (4) recording disclosures, and (5) speculation.

Recording Conception of an Invention

When you have that flash of insight, that "Aha!" experience that leads you to invent something, write your thoughts into the notebook without delay. Describe the factual circumstances surrounding your invention. Note whether conversations, correspondence, journal articles, or perhaps a chance observation contributed to your invention. Give credit where credit is due to others who contributed to the conception of the invention. However, avoid using a phrase such as, "It is obvious from prior work that...." If your notebook is not at hand and you will be unable to write in your notebook for several hours or longer, write your thoughts on a piece of good quality paper with permanent ink, and sign and date the paper. Later, when you sit down to put the thoughts into your notebook, you will have already documented the circumstances under which you made the invention.

There are two schools of thought about putting such notes into the notebook. One school advises that putting any small pieces of paper into the notebook raises suspicion that the note may be out of chronological sequence, that it was inserted into a place in the notebook for selfish reasons (i.e., to establish an early date of invention). (I have actually heard of a lawyer who demanded in court that a scientist unglue a pasted-in note to prove that a previous entry was not being concealed.) This school advocates that the best and only way to document conception of an invention in the notebook is to make a regular entry, handwritten directly in the notebook, that is signed, dated, and witnessed at the earliest possible convenience of the inventor. Pieces of paper taped to notebook pages should never be relied upon to prove anything.

The other school of thought advises that because an inventor's notes are deemed by the court to be self-serving, and because any testimony must be corroborated by a disinterested witness, pieces of paper may be inserted into the notebook if they are properly witnessed. The note itself should be signed and dated by the inventor, and the witness can sign and date the note by writing across the edge of the note and onto the notebook page to which it is attached. (*See* Figure 7.1.)

A few hours difference in the time between actual conception of an invention and the time that it is written into the notebook is unlikely to make a significant difference in protecting your priority of invention. Delays of several days, on the other hand, may very well be significant. Therefore, if you will have access to your notebook within the day on which an invention is conceived, do the following: Write your thoughts onto whatever paper is handy as soon as the invention is conceived. This action serves to fix the time, date, place, and circumstances under which the invention was actually conceived. Then, transcribe your notes into the notebook as soon as possible and have them witnessed as soon as you have finished writing the details in the notebook. This action ensures that a witness can corroborate the date on which the invention was conceived. If you will not have access to your notebook for several days, try to find a competent witness who can sign and date your little note. Later, when you do attach the note to the notebook, make sure the attachment is properly witnessed at that time. Remember, you will only be able to prove the earliest date for which you have a corroborating witness. An unwitnessed written description of a new invention is little better than no description at all when proving the date of conception.

Continuity

If your patent is the object of an interference proceeding, you will need records to substantiate the date of conception and your diligence in reduction to practice. Proving diligence can be very difficult, but the notebook can help perhaps more than any other device. To demonstrate diligence, you must show that you worked continuously on your invention, that you did not set it aside or abandon it. In highly competitive fields, the inventor should take special pains to show that work to make the first working version of the invention was continuous.

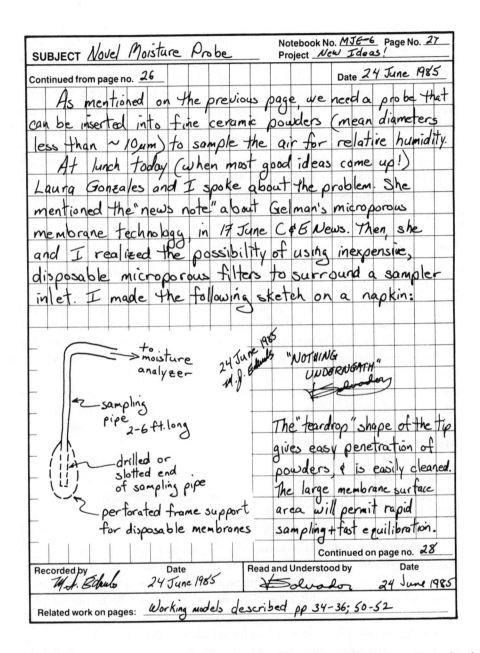

Figure 7.1 An example of how to attach a small note to a notebook page. The archival quality mending tape described in Chapter 2 was used and is virtually invisible. Note the signatures of both the author and a witness. The witness is required to corroborate when the note was entered into the notebook. The notation "Nothing underneath" emphasizes that the note was not back-dated to suggest an earlier-than-true date.

To show that you were indeed diligent, do the following. From the time when an invention is conceived, make an entry in your notebook on the subject every day. If you did not work on the invention for several days, explain why not. For example, if you were waiting for samples to come back from another lab or if parts for your apparatus were being made in the machine shop, then say so. If you thought about your invention but did not make any significant progress, then record what you thought about. If you were away from work due to illness or vacation, then make that clear. Have a competent witness sign your notebook as the work progresses, and let no more than a few days elapse before subsequent witnessing. Remember that you are trying to show in your notebook that you were striving diligently to make your invention workable.

Witnessing

If your patent application becomes the object of an interference proceeding or other legal challenge, you may be called to testify about the circumstances surrounding the conception of your invention and your attempts to reduce the invention to practice. You may be questioned during the legal process called *discovery*, and you may be asked to give a sworn deposition on the subject of your invention. Your notebook (or a facsimile of part of it) will be introduced as evidence to support your testimony. However, the testimony of an inventor is deemed to be self-serving and must be corroborated by a witness who is not a coinventor. Who can serve as a witness? Current interpretation of the law requires that the witness be someone who is technically competent to have understood your invention at the time it was conceived.

As a guide for selecting a competent witness, consider the following. The specification for the invention submitted in the patent application must be "... in such full, clear, concise, and exact terms as to enable any person skilled in the art to which it pertains...to make and use the same" (4).

S. B. Williams stated (5), "A person having ordinary skill in the art is a person who understands and is knowledgeable about prior inventions in the field to which the invention relates." If you use this definition as a guide for finding a competent witness, two other considerations are important. Any witness you choose must be trustworthy because he or she may have to testify or provide depositions on your behalf. For the same reasons, choose a witness who will be available for the next few years in case his or her testimony is needed.

J. H. Wills (6) recounts the case of Scharmann vs. Kassel (7) in which the witnesses' testimony was found by the court to be even stronger proof of conception than the inventor's own somewhat careless records: "The forthright testimony which convinced this judge is found in the example of one witness who said, 'I distinctly recall that Dr. Kassel on May 6, 1938 disclosed to me sketches of the design of the apparatus I had previously described.' ...the court said, 'testimony of both those witnesses...stands as positive and forthright corroboration....'"

More recently, O. M. Behr (8) gave a taste of just how complicated litigation can be when new chemical compositions of matter are involved: "In considering the witnessing of a conception, it is necessary for the witness to understand what is being proposed. This is not the case in corroboration of reduction to practice. All that is required is that the witness be able to testify to what was actually done."

Behr goes on to cite (9) the example of a case in which a new microbiological preparation was conceived: the substance was synthesized by the combined efforts of several people, and its nature was confirmed by still others. Most of the records involved in this case were unwitnessed, but "... the court held that while there was in fact a gap in the actual corroboration of the synthesis of the compound, there was such a cohesive web of circumstance around this work that it was reasonable to assume the truth of the matters alleged by [the inventor]."

This story suggests that, although circumstances might lead to winning the case if "the truth" is on your side, properly witnessed notes are preferable. In recording the research surrounding the invention of a new composition of matter, you must know and demonstrate the composition of reagents, starting materials, intermediate compounds, and the products of your work. Witnesses must be able to provide convincing testimony, backed by analytical and other records.

Often during pretrial gathering of evidence, one side feels that it has sufficient evidence in its favor to suggest a settlement with the opponents before the case goes to trial. Well-kept records can no doubt provide your company with such evidence and may save your company considerable additional legal fees, time, and embarrassment.

This discussion suggests that some of your colleagues and even your supervisor may not really be the best witnesses for your notebook. Do not consider someone from your clerical staff or a local notary public as a competent witness. Technicians working under your supervision may be acceptable witnesses if they are extremely knowledgeable about the invention and will definitely not be named as coinventors. A person who is reading a notebook as a witness may have an idea that may lead to coinventorship. In that event, simply have subsequent entries witnessed by another person.

The witness must make clear what he or she is witnessing. Has the witness simply read the entry or actually observed an experiment? This matter can be made clear by having the witness's signature preceded by one of the following phrases: "This page read and understood by _____ (signature, date)" or "Pages ___ through ___ read and understood by _____ (signature, date)." If appropriate, one of the following phrases may be used: "Experiment observed by _____ (signature, date)," "Operation of this apparatus observed by _____ (signature, date)," "Material tested in my presence _____ (signature, date)," or "Samples synthesized in my presence _____ (signature, date)."

If, on the other hand, a witness simply signs a notebook page "Witnessed by me _____ (signature, date)," doubt can exist as to exactly what was witnessed. Was the inventor's work, the notebook entry, or perhaps just the inventor's signature witnessed?

In the field of novel compositions of matter, new chemical processes, and the like, the work should be repeated as soon as possible after the invention has been made. The work must be repeated by someone following the inventor's description of how to make or use the invention. The work can be repeated by anyone as long as that person can be reliably expected to serve as a witness if needed. Such a person should write a description of the repeated experiment into the inventor's notebook and sign and date the entry. The experiment may be observed by several people who also serve as witnesses. This procedure corroborates the inventor's work.

If you are asked to be a witness to a colleague's notebook, be sure that you indeed understand the subject of the passage that you are witnessing. Ask the writer questions if anything is unclear and do not be afraid to suggest that the writer expand an explanation, drawing, or description to improve clarity and to remove any ambiguity. You may be called to testify about the notes or work that you witnessed; once you have been "raked over the coals" by opposing counsel you will appreciate the importance of being a competent witness.

If you observe a demonstration as a witness, be sure to record that fact in either your own notebook or the notebook of the inventor. Your oral testimony alone will not be considered as reliable as your notebook entries, especially if significant time has elapsed since the observations were made.

As a summary of the subject of witnessing, the following is from a Patent Office publication (10): "Your priority right against anyone else who makes the same invention independently cannot be sustained except by testimony of someone else who corroborates your own testimony as to all important facts, such as conception of the invention, diligence, and the success of any tests you may have made."

In other words, all of your hard labor may be for naught if proper witnessing is neglected.

Recording Disclosures to Others

To protect yourself against theft of your invention, keep knowledge of the invention confined to a close group of people. If you discuss the invention with others, make a notebook entry describing to whom the invention was disclosed and under what circumstances. Note if the other person made a significant contribution such as a suggestion to modify or improve the invention, or possible uses. If an invention was jointly conceived, then make that point clear. Often, you as the inventor will come up with a good idea after putting together information from various sources. Note dates, times, places, and events of significance that might have some bearing on establishing your priority as the inventor.

If you submit a written description of your invention to anyone (i.e., a colleague at another institution or a journal), note when you sent the description.

Cross-reference your notebook to your copy of the correspondence. If you make an oral presentation about your invention, note that fact. If discussion ensues, note the general theme of the discussion and name the people present, if feasible. If you attend a staff meeting to discuss your invention, take along your notebook and make your notes directly into the notebook rather than write them on a tablet for later transcription into the notebook. This practice will (1) save you the time of transcribing the notes later, (2) reduce the chances for errors or inaccuracies when you later try to recall elements of the discussions, and (3) provide a witness to your notes—another person who was at the meeting can witness your notes as an accurate account of the discussions.

Speculation

The notebook is mainly a collection of facts. Speculation, however, is very appropriate in one special circumstance: when you are suggesting novel uses for a new invention. This section is important for two reasons. First, an invention must be useful in order to be patentable. Second, a new use for an established material can be patentable. For example, the use of aspartame as a food sweetener is covered by a patent, although aspartame itself is a naturally occurring substance and is therefore unpatentable as a new composition of matter.

Wild speculation (or brainstorming) on possible uses for your new invention is appropriate. Remember, the experimental sections of your notes should contain complete, factual accounts of your work, including positive and negative results. However, avoid suggesting that your invention is unsuitable for certain uses. If you record that you see no practical uses for your invention, you may be seen by the court as having abandoned your invention.

Protection of Trade Secrets

Most, if not all, of your notes are the proprietary property of your employer and are subject to laws on trade secrets. (This subject is discussed in Chapter 3.) Any invention should be discussed only with people authorized to learn about the invention, and notebook pages should not be copied or distributed without permission from management. The notebook should be safely and securely stored (*see* Chapter 4) to minimize the chances of loss or theft.

Literature Cited

1. Maynard, J. T. *Understanding Chemical Patents*; American Chemical Society: Washington, D.C., 1978; p. 30.
2. Arnold, T. and Vaden, F. S. *Invention Protection for Practicing Engineers*; Barnes and Noble: New York, 1971; p. 136.
3. "General Information Concerning Patents;" U. S. Department of Commerce Patent and Trademark Office, Feb. 1983, p.23.
4. 35 U. S. Code, Section 112.
5. Williams, S. B.; *Science* 1984, 225(4657), 18–23.

6. Wills, J. H.; *J. Chem. Ed.* 1953, *30(8)*, 407–410.
7. Scharmann vs. Kassel 636 O.G. 391 (1950).
8. Behr, O. M.; *CHEMTECH* 1982, August, p. 490.
9. Berges vs. Gutstein, Kaplan, and Granatak 205 *U.S. Patent Quarterly* 1980, *691*.
10. "Patents and Inventions: An Information Aid for Inventors;" U. S. Department of Commerce Patent and Trademark Office, Aug. 1977, p. 4.

The Electronic Notebook

Computers are performing many of the notekeeping functions that the research scientist traditionally did with pen and paper. These tasks include sample documentation, instrument calibration, raw data collection, and data reduction. In considering the role of computers as *electronic notebooks*, we should make a distinction between the laboratory that does mostly routine work (e.g., chemical analysis or quality control testing) and a research lab in which an experiment is rarely repeated. Semiautomated and fully automated systems for sample preparation, analysis, and report preparation can drastically reduce costs in the labor-intensive analytical lab. The analyst's job can be made safer and easier by such automation, and the possibility of making mistakes is minimized. However, the analyst rarely needs to sit down and write notes the way a research scientist does. In research laboratories, expository writing (e.g., describing an invention, stating a hypothesis, or sketching chemical structures and reaction mechanisms) is still done in a handwritten notebook.

Most good notekeeping practices developed for handwritten records can be used with the newer systems. In addition, the power of today's electronic systems suggests several areas in which these electronic systems can be superior.

The shift in reliance from the traditional paper-based system to electronic "paperless" systems raises many hopes and many questions. This chapter has two purposes: (1) to compare the advantages and disadvantages of both electronic and paper-based notekeeping systems, and (2) to discuss the features in an electronic system that will make it truly worthy of the title, "the electronic notebook."

Recent reviews have been published on the subject of laboratory computer networks (1) and workstations in the laboratory (2). The proceedings of a symposium devoted to laboratory automation have also been published (3). These three references give a good overview of the state of the art of laboratory computers.

0906–5/85/0113/$06.00/1
©1985 American Chemical Society

On the following pages is a summary of the advantages and disadvantages of traditional, handwritten notekeeping systems and computer-based systems. Because automated data acquisition is already common in testing laboratories, the following comparison focuses more on the type of notebook entries that are made by scientists in research laboratories. The purpose of this comparison is to stimulate discussion among users and producers of electronic notebook systems regarding desirable features of future systems.

Comparison of Electronic and Handwritten Notebooks

TASK—Writing and Reading the Notebook

THE HANDWRITTEN NOTEBOOK   *Advantages*   Everyone knows how to write notes by hand; only paper and a pen are required. The notebook is always ready and there is no need to gain access or log on. Drawings can be made to fit exactly into the available space. Figures, tables, and graphs can be easily placed together on the same page, if wanted. Pages can be quickly flipped and side-by-side comparisons of notes can be made easily.

Disadvantages   Sloppy handwriting is often hard to read and data can be misread. Data can be lost if something is spilled on the page. Pages can be accidentally torn and notes can be obliterated by solvents.

THE ELECTRONIC NOTEBOOK   *Advantages*   Output on visual displays is highly legible. Input can be much faster than handwriting if the author is a good typist. Automated data acquisition systems can provide unattended, virtually error-free collection of raw data from instruments. Integration of text, graphs, or data tables is possible. Automated searching can be used to quickly find references to any particular topic. Systems with graphics capability can produce high-quality drawings. Several people can have simultaneous access to the same data so that supervisors can check up-to-the-minute progress of work, even from distant research centers. Automated page readers can be used to turn handwritten notes into electronic notes that can be combined and stored with other data. Voice synthesizers can speak the words to visually impaired persons. Manufacturers claim that error detection schemes can produce perfectly accurate records of the collected data during initial writing onto optical storage media.

Disadvantages   Typing skill is required to enter data efficiently. The user must learn how to use the particular system. (By comparison, skill in handwritten notekeeping can be taken to any job.) The computer or a terminal must be in the work area and must be accessed (i.e., it must be turned on, programs must be loaded, and the user must be logged on). The system may be unavailable when needed. Problems with the computer can cause information to be lost during the process of data entry, and the information must be reentered. Long sessions at the console require well-designed computer furniture and appropriate lighting to minimize fatigue.

TASK—*Transporting the Notebook*

THE HANDWRITTEN NOTEBOOK   *Advantages*   A bound notebook is small, lightweight, easy to carry anywhere (i.e., to the conference room, lab, pilot plant, office, home, or into the field). Use is not limited by the lack of electricity or air conditioning, or by the presence of high humidity or dust. Sturdy case-bound notebooks can be handled quite roughly with little consequent damage.

Disadvantages   Notebooks can easily be misplaced when moved about. If not properly labelled, notes may be lost.

THE ELECTRONIC NOTEBOOK   *Advantages*   The location of the computer or a terminal is always known. (A terminal or small computer cannot be easily misplaced!) Some small computers can be taken anywhere you can take a book, although computing power is relatively limited. Large volumes of information on optical or magnetic storage media can be sent to remote locations at relatively low cost. Data in electronic notebooks can be exchanged rapidly over long distances, if necessary.

Disadvantages   Most computers currently useful for notekeeping are not portable and must be used in dry, temperate atmospheres. A source of power is always needed. Magnetic storage media are susceptible to corruption by stray magnetic fields, dirt, moisture, and rough handling encountered during travel.

TASK—*Ensuring Security of the Notes*

THE HANDWRITTEN NOTEBOOK   *Advantages*   Notebooks can simply be locked up for safekeeping in a convenient place such as a desk drawer. The notebook's author, as well as others, has easy access. Because of witnessing requirements for a properly kept notebook, collusion would be needed to forge entries or to change dates that will stand up in court. Data generally cannot be obliterated or altered without leaving some trace of the alteration.

Disadvantages   Notebook pages can be easily copied and distributed. Data can be intentionally or accidentally obliterated. Notebooks are susceptible to several forms of damage (*see* the following section on storage). Notebooks can be taken from the premises without permission of management.

THE ELECTRONIC NOTEBOOK   *Advantages*   Access can be strictly limited through a combination of user accounts, passwords, and other security measures. Several levels of access can be established such as "read only," or "read and add but cannot alter." Entries can be date- and time-stamped by the system.

Disadvantages   No commercially available system is capable of proving who was the author of a particular notebook passage, and no provision is allowed for witnessing. Data on magnetic media can be overwritten by the notebook author because there is no "write-once-only" system. (Optical disk technology will perhaps provide this capability.) Chronology of entries is subject to question if files are not strictly sequential and unalterable.

TASK—Storage of Completed Notebooks

THE HANDWRITTEN NOTEBOOK   *Advantages*   Notes written with archival quality ink on permanent-durable paper will last indefinitely. Storage is relatively cheap and requires controlling the temperature, humidity, and physical security.

Disadvantages   Paper-based records are susceptible to attack, by fire, bacteria, insects, or rodents, for example. If low quality paper was used or if the notebooks were improperly stored, the pages may be extremely discolored, fragile, or illegible after only 10–20 years.

THE ELECTRONIC NOTEBOOK   *Advantages*   High-density storage media can save a tremendous amount of space over paper-based records. Optical disk technology can accommodate more than 100,000 pages of text on a single, thin disk approximately 14 in. in diameter (as of April 1985). Optical disks need no routine maintenance while in storage. Algorithms and equipment used to read the optical disks are designed to correct random errors. Combined with the stability of the medium, this system is said to be capable of limiting errors to less than 1 bit in 10^{13} during the projected disk life of 10 years.

Disadvantages   Magnetic tapes must be "exercised" (rewound and cleaned) at least once every year to avoid physical "blocking" of the tape (adjacent layers of tape sticking together) and to minimize migration of magnetic domains from one layer of tape to the next. Tapes also need to be read through certifiers or evaluators every few years to locate and correct errors that have occurred during storage. Algorithms and equipment needed to read the electronically stored data must be available and in working condition for as long as the records might be useful, several decades at least. Storage facility requirements are similar to those for paper-based records, including temperature and humidity control, and cleanliness. (*See* Reference 4 for a complete discussion of the care and handling of magnetic media.)

Discussion

Electronic systems suitable for notekeeping can be found in sizes ranging from desktop microcomputers to laboratory mainframe systems. Several vendors now supply software meant to be used for data acquisition, data reduction, report preparation, and graphics. (*See* the list at the end of this chapter.) However, none of these systems contains all of the features desirable in a true electronic notebook.

What is missing? First, computer manufacturers and software vendors seem to be primarily concerned with data transmission speed, computing power, and interfacing with peripherals. These features are important in a highly competitive field where speed of data acquisition and calculation influence the purchaser. Security and ease of use are also very important for the note writer. For example, a scientist can sit down at a desk, pick up a notebook and pen, and begin to write without worrying about passwords, user accounts, or whether the system is up or down. This sort of effortless user recognition by the system should eventually be part of electronic notekeeping systems.

Another shortcoming of today's systems is that their bulk and delicate construction mean that they cannot be used in many environments into which notebooks can be carried. Future electronic notebook terminals will have to be robust and capable of operating in dusty or wet environments.

The electronic notebook should be able to do much more than simply replace the handwritten notebook. For example, the electronic system should be capable of stamping the time and date for each entry in such a way that when the entry was made can be proven. In addition to date-stamping, the electronic system should have a provision for the electronic "signatures" of the author and witnesses. How might this provision be accomplished? Many systems are commercially available, or under development, that are based on various individual characteristics, such as fingerprints or palmprints, retinal blood vessel maps, voice recognition, and sensing the pressure and timing of handwritten signatures on a digitizing tablet. One, or a combination of several, of these systems might become a standard method of verifying the authenticity of a signature.

How can the time and date of a notebook entry be proven without relying on a human witness? When a note writer creates an entry of significance that is to be "electronically witnessed," the notebook entry could be electronically transmitted to a computer (perhaps maintained by the U.S. National Bureau of Standards) that creates a unique *checksum* out of the notebook entry by using an algorithm containing the date and time. The checksum is then immediately transmitted back to the electronic notebook and recorded. When proving the date of an entry becomes necessary (for example, in a patent dispute), a copy of the notebook entry along with the recorded checksum can be sent to the government agency for verification of the date on which the entry was made. The algorithm for creating the checksum can be kept secret and changed periodically to minimize the chance that someone will be able to forge the date-stamp. This method has the advantage that, whereas the date is verified by a disinterested third party, such as a government agency, that agency does not have the burden of actually recording and preserving all the thousands of notebook entries it might receive each month.

Another area that needs improvement is to make electronic storage media of high storage density that have archival stability. Such storage media should be *human readable* as well. This last point needs some clarification. In several instances, computer-readable documents have been preserved but the algorithms for reading and deciphering the information on the magnetic tapes have been lost (5). The tapes are therefore useless. The electronic storage medium must be either directly readable by humans (e.g., through a microscope, written in standard code, or like an ASCII punched paper tape) or set up so that every storage device is somehow labelled with instructions for reading it. Such media should not be device dependent or encrypted in such a way as to make them unfathomable if the decoding algorithm is lost. By way of comparison, realize that any old notebook can be simply picked up and read (as long as the language is known to the reader). So it should be with electronic devices.

What will the future of electronic notekeeping really be like? For the next several decades, more and more primary data will be recorded directly by electronic means. More and more data reduction will be done by computer with less and less handwriting. The expository writing that the average research scientist does today will still be done in the bound notebook, signed and witnessed. Eventually, however, this sort of writing will be replaced by text entry through a keyboard or by speaking through voice-recognition systems.

Validation

How do we validate laboratory computer systems, that is, how do we show that the computer is collecting and recording data without error? Put another way, how do we know that the computer is doing what it is supposed to do, and that it will continue to do so?

This subject was addressed recently at an ACS National Meeting (6). A session was devoted to discussing validation of computer systems and computer records in regulated laboratories. Most of the discussion concerned work performed in laboratories subject to the U.S. Food and Drug Administration's *Good Laboratory Practices*, although many of the practices discussed could be used to advantage in other types of laboratories. According to the participants, at a minimum computers should be programmed to check for nonsense input (e.g., instrument readings beyond working ranges) thus performing one of the analyst's prime functions. More subtle errors such as dropping a bit in data transmission can be handled by error-checking schemes. Recovery after planned shut-downs or after unexpected failures must include rechecking the validity of the system.

Validation must be done on a case-by-case basis; each lab manager must know the system's hardware and software limitations. Users must be aware of potential weak points in the particular system under question. Tests must be run to demonstrate that the system does indeed produce the expected results when running under known conditions. The system must be monitored to ensure that data collection and recording are being performed correctly.

Summary

- Automated data collection and recording is already well established in testing laboratories. In research laboratories, however, routine expository writing is still done in handwritten notebooks and will continue to be so for some years to come. Professors and students who are not concerned about witnessing and validation of data as required in many industrial laboratories will find that existing electronic notebook systems are quite adequate for most purposes.

• Electronic notebooks have many potential advantages over handwritten systems, including rapid data entry; security of data; sharing and high-speed transmission of data to distant locations; rapid searching and retrieval of specific information; manipulation of data into graphs, charts, and drawings; and high-density storage for large volumes of raw data.

• To realize these potential advantages and to replace handwritten notebooks in industrial laboratories, electronic notebook systems must evolve beyond their present, limited powers. Electronic notebook manufacturers might see the future divided into several short-term and long-term goals:

Short-term—To be useful for research notekeeping, existing word-processing systems for desktop microcomputers can be modified to time- and date-stamp entries, and to provide controlled levels of access. A common format for electronic data transmission will be necessary to move and share data among different workstations.

Long-term—The electronic notebook of the future should have the following features:

- Immediate, passive recognition of the user.
- Data writing by voice.
- Unforgeable and unalterable time- and date-stamping of entries.
- Provisions for electronic signatures of witnesses.
- Small, lightweight, portable; insensitive to harsh environments.
- High-density storage media with archival stability, human-readable format.

Appendix. Microcomputer Resources

The following companies (among others) sell microcomputer software to acquire laboratory data that can then be passed to database, spreadsheet, or other programs for computation and plotting:

Bolt Beranek and Newman, Inc.
10 Moulton Street
Cambridge, MA 02238
(617) 491-1850

Product: "RS-1" for Digital Equipment Corporation computers including VAX, PDP-11, and Professional 350. Also for IBM Personal Computer XT and AT.

Laboratory Technologies Corporation
255 Ballardville Street
Wilmington, MA 01887
(617) 657-5400

Product: "Labtech Notebook" for IBM Personal Computers XT and AT, COMPAQ computers, Data General One computers, and other microcomputers.

MacMillan Software Company
866 Third Avenue
New York, NY 10022

Product: "ASYST" integrated software for data aquisition, analysis, and presentation for the IBM PC, XT, and AT.

Comprehensive laboratory information management systems (LIMS) are available through many laboratory instrument makers including the following:

Computer Inquiry Systems, Inc.
Division of Beckman Instruments
160 Hopper Avenue
Waldwick, NJ 07463
(201) 444-8900

Perkin-Elmer Corporation
Main Avenue (MS-12)
Norwalk, CT 06856
(203) 762-1000

Hewlett-Packard Company
Laboratory Automation Systems
3000 Hanover Street
Palo Alto, CA 94304-1181

Digital Equipment Corporation
Laboratory Products Group
One Iron Way
Marlborogh, MA 01752
(617) 897-5111

Varian Instrument Group
Laboratory Data Systems
2700 Mitchel Drive
Walnut Creek, CA 94598
(415) 939-2400

The April 26, 1985, issue of AAAS *Science*, Volume 228, No. 4698, contained 11 comprehensive review articles describing the status of computers in research and science education.

Literature Cited

1. "Laboratory Computer Networks" *Anal. Chem.* 1984, *56(3)*, 408A.
2. Dessy, R. E. *Anal. Chem.* 1985, *57(1)* 77A.
3. *Computers in the Laboratory*; Liscouski, Joseph, Ed.; ACS SYMPOSIUM SERIES No. 265; American Chemical Society: Washington, D.C., 1984.
4. Geller, S. B. "Care and Handling of Computer Magnetic Storage Media;" U. S. National Bureau of Standards Special Publication 500–101, June 1983.
5. "Report of the Committee on the Records of Government;" May, E. R., Chairman; March 1985; available from The Council on Library Resources, Washington, DC 20036.
6. "Symposium on Government Regulations on Computers in Laboratories;" sponsored by the ACS Division of Computers in Chemistry at the 188th ACS National Meeting, Philadelphia, 27 August 1984.

Appendix A

Some Suggestions for Teaching Laboratory Notekeeping

Why Teach Laboratory Notekeeping?

It is not unfair to ask, "Why teach laboratory notekeeping at all in school? With so much for the student to learn about science, shouldn't notekeeping be learned on the job? Surely each company has its own rigorous guidelines for notekeeping that are different from other employers."

To answer those questions, consider the following points that were mentioned briefly in Chapter 1. Notekeeping is an acquired skill that can be of tremendous benefit in almost any career. Proper notekeeping should be a matter of habit, not a chore and, therefore, should be developed in the student at an early age, certainly before the student begins a scientific career. Good communication skills are important in every job and notewriting is a fundamental personal communication skill. Regrettably, the ACS Committee on Professional Training (ACS–CPT) recently stated (1), "Employers of chemists report to the Committee that a large fraction of baccalaureate chemists write and speak poorly." The Committee went on to recommend, "Laboratory work should give students hands-on knowledge of chemistry and the self-confidence and competence to... keep neat, complete experimental records....Frequent exercises in writing and speaking, critically evaluated by the chemistry faculty, are an essential part of a sound program in chemistry."

Most employers have specific rules for notekeeping. Although each set of instructions may be worded a little differently, all have approximately the same basic requirements and can therefore be taught at all schools. Specific notekeeping duties will of course be different at different companies; notekeeping duties within a company will vary at different jobs. However, the principles of good notekeeping for working efficiently and for protecting inventions are the same at any job. An understanding of these principles and knowledge of how to

0906–5/85/0121/$06.00/1

implement them is precisely the sort of skill that students should have before they begin job-hunting.

A prospective employer may ask the student to make a presentation on the subject of the student's research work. Such a session gives the company a chance to learn about the student's oral communication skills. The student gets quizzed to find out just how well he or she knows the subject matter. A company will sometimes request a copy of the student's thesis or dissertation, if it relates to the subject of the pending job. A wise corporate recruiter will also ask to look at the student's notebooks—just a glance can yield an excellent insight into the student's attitudes about neatness, thoroughness, attention to details, clarity of thought, and clarity of expression. Students should be proud to show their notebooks to recruiters as an example of an important and well-developed skill.

When to Teach Notekeeping?

What is the best age for students to learn notekeeping? Knowledge about laboratory notekeeping should be a three-step progression as the student advances: (1) The grammar school student begins to understand that a scientist makes a record of every experiment. (2) The high school student learns clear expression. (3) The college student learns the detailed principles of proper notekeeping with an eye toward getting and holding a job. If teachers at all levels can agree on this scheme, students will indeed develop proper notekeeping as a matter of habit. They will work more efficiently in graduate school, and they will be more valuable as employees. If the employer has to spend less time teaching fundamentals such as notekeeping, more time can be devoted to solution of the actual scientific problem at hand.

To make notekeeping habitual, the student should be taking notes each time work is done in the lab or field. Students should feel comfortable with a notebook in hand at the start of every experiment, and they should feel just a bit naked without a notebook. In the same way in which they get used to taking notes in a lecture class, students should expect to write notes while doing an experiment. Too often, students look at the lab period simply as a chance to get away from rigorous lecture sessions. In college, teaching assistants often "reward" students who quickly finish the lab sessions by allowing them to leave as soon as the experiments are done; notemaking and writing lab reports are postponed until some later time. Such a practice reinforces several wrong attitudes including, "finishing the work fast is more important than doing the job carefully" and "writing up the work is a necessary evil, better left for later."

Some General Comments

Experimental science (the subject of most laboratory notekeeping) is descriptive science; communicating the results of scientific investigations requires strong skills of description. Too often a *written* description is very different than the *spoken* description. Much writing could be improved if the writer would simply say the words out loud, write down what sounds good, and reject phrases and words that sound awkward or stilted.

The chemistry teacher has a real challenge in teaching his or her subject matter. Chemistry is considered by many otherwise bright students to be an abstract and difficult subject. After all, who can actually show the student an atom? In contrast, the basic principles of physics (mechanics) are often taught by using ropes, pulleys, and hockey pucks on frictionless air tables. The actions and reactions of objects are easily seen and easily grasped. Biology students study animals and plants; students can actually see the subjects of their studies. Geology students can handle rocks; astronomy students can see the stars. But chemistry teachers must try to explain to the young chemistry student why a molecule of bromine produces an acrid, intensely colored liquid and a molecule of oxygen produces a colorless, odorless gas that sustains life. They must explain how sodium, a highly reactive metal, can combine with chlorine, a toxic gas, to produce common table salt. The fundamental subjects of the chemistry student's studies, atoms and molecules, are inaccessible to direct observation using the five senses. As a matter of philosophy, students should understand that chemists must perform experiments and record and interpret the results of these experiments to understand chemistry. Merely filling in the blanks of an answer sheet is unsatisfying and, in the long run, unfair to the student.

That scientific writing must be done in the third person, by using the passive voice, is a myth: "It was observed that the mixture changed color after a few minutes had elapsed." According to this myth, such a tone lends an air of professionalism or objectivity to the writing. In fact, most major guidelines for authors of scientific publications stress that authors use the active voice in the first person, when appropriate, to express an opinion or to state a fact. Much writing could be improved simply by using the active voice, first person: "I saw the mixture turn blue after 10 min." leaves no doubt as to who did the work, what happened, and when it hapened. This tone is vigorous, concise, and to the point. It is the tone of choice for basic descriptive writing, and it is the tone of choice for laboratory notekeeping.

When students become employees, they will have to take responsibility for their own opinions and recommendations: "Based on my experiments, I recommend that we abandon this project." They should be unafraid to say "I" or "we" when appropriate. The lab notebook is as good a place as any for getting used to shouldering the burdens of responsibility.

We can, however, make the distinction between notewriting and the preparation of a scientific article. The first criterion applied to judging papers submitted for publication in scholarly journals is readability. A combination of active and passive sentence constructions can help to develop a good rhythm and can make the paper more readable. Continuous use of the active voice results in tiresome, choppy sentences. Continuous use of the passive voice can be confusing at worst; at best it lulls the reader to sleep. The active voice reminds you, the reader, that the author is talking to you.

Teaching Notekeeping in Grammar School

This time is such a marvelous one for budding young scientists; there is something of interest for every one. They go down to the pond and get a glass of pond water; they place a drop of this water on a microscope slide and observe the drop under the microscope: Hundreds of wriggling creatures never before imagined come into view. The students collect leaves and compare them for shape, color, texture, and odor. Student stargazers learn about patterns of stars in the sky, and, if their eyes are sharp, they see that stars have different colors and that many different types of objects are in the night sky. Young rock hounds gather specimens for examination of color, hardness, streak, and luster.

Why not write notes about these experiences? Each student should be issued a small notebook to keep all year. For each experiment or specimen-gathering trip, the students should write down something. What they write isn't important, but they should get into the habit of simply writing down something about what they did, saw, smelled, felt, or thought. "Describe what you saw" or "tell me what you did" are the only instructions that the teacher need give.

At such an early age, rigor and rules are secondary to making the student feel comfortable and become accustomed to recordkeeping in a lab notebook. Each student can be issued a small bound notebook that can be called a lab notebook, journal, or diary. The point is that the young student should be motivated to write down as much as possible. More detailed instructions will come in later years.

The following is an exercise to get young students to improve their powers of observation and description: Put an assortment of laboratory objects into a box and ask each student to take one object out of the box. The student must then describe the object well enough (without drawing a picture!) so that other students can recognize the object from the written description. Use objects of different materials and textures (e.g., rubber, metal, glass, plastic, wax, or stone) some of which can be quite similar except for small differences (e.g., left- versus right-handed!). Remind the students to note all the properties they can, including color, texture, odor, size, shape, hardness, or sound when tapped. Ask the students to speculate on possible uses for the objects. When the students have finished, put all the objects on a table and collect the written descriptions. Read them, one at a time, and have the students figure out which object is being described. You can then explain the proper names of the objects and their actual uses. Some objects could be connected, such as lab glassware, so that they must be assembled for use. If you now let the students add drawings so that they can better describe the objects, they will also see how important a good drawing can be.

Teaching Notekeeping in High School

In high school, the emphasis should be on clarity of descriptions: Can your classmates understand what you wrote? Are you using good grammar and good spelling? Is your writing legible? Are your drawings understandable? If grammar school piques students' interest in science, the high school years expose them to the nuances and differences among the disciplines. This early exposure often leads to a career choice before the students begin college.

Shakhashiri points out (3), "one of the purposes of demonstrations is to increase the students' ability to make observations." Lecture demonstrations (also called lecture experiments) performed by the teacher can be used to improve notekeeping skills. Classic demonstrations such as burning a strip of magnesium in air or reacting a small piece of sodium with water can be performed while students observe. At appropriate times, the demonstrator can pause to let the students write a description in their notebooks of what they have seen. These descriptions should just be simple narratives. Rather than the usual sort of classroom notes, the students should write their notes as if they were reading them to someone who wasn't present, to tell a story. (*See* Chapter 6 for examples of well-written notebook passages.)

High school students should begin to understand that careful layout of the notebook (e.g., organized tables or labelled drawings) will help when the time comes to interpret experimental data. Students who make projects for local science fairs should be especially encouraged to keep good notes of their studies.

At the Eighth Biennial Conference on Chemical Education (2), Tony Neidig conducted a workshop on developing experiments for chemistry classes. He discussed the differences among three types of lab sessions: a laboratory exercise (e.g., distillation of wood splinters), an experiment (e.g., qualitative analysis of a mixture of unknowns), and an investigation (e.g., the effect of acid rain on different materials). In a laboratory exercise, the teacher knows the answer, the student knows the answer, and the session is likely to be boring. In an experiment, the teacher knows the answer, the student usually doesn't, but the mechanical aspects of the session still can be boring. In an investigation, neither the teacher or the student knows the answer and the session is bound to be interesting!

Part of performing any investigation is keeping notes so that the results can be analyzed and conclusions can be drawn. Many laboratory exercises will not lend themselves to teaching notekeeping because the students know the supposed outcome and will mechanically go about the work. The teacher's exhortation to explore and to discover things during the course of a laboratory investigation requires good notekeeping and thereby teaches good notekeeping.

The participants at this workshop agreed that designers of experiments should keep in mind cost, safety, interest, challenge, simplicity, and gradeability. Almost any experiment can be modified to include notekeeping, and such a change will meet these several criteria.

Teaching Notekeeping in College

The freshman year of college is perhaps the worst time to try to teach laboratory notekeeping, especially in the large classes at major universities with thousands of students and dozens of graduate teaching assistants. For students, it is the year of moving away from home, of breaking old bonds and making new ones, of struggling to stay afloat in a sea of new experiences. Most students taking introductory chemistry courses will not become chemists, although some will end up in scientific disciplines. Many must take chemistry as a prerequisite for courses in nursing, geology, biology, and other subjects. The sheer numbers of freshman taking chemistry precludes adequate critiquing and discussion of notekeeping with the teacher. Most graduate teaching assistants have neither the experience or the ability to properly teach notekeeping. Instruction should be reserved until the sophomore and later years, when classes are smaller and more attention can be given to the individual student.

Upper level science majors should certainly be taught the details of proper notekeeping. All of the subjects in this book are appropriate for the student who will be entering industrial or graduate research. In light of the ACS–CPT report, students should realize that good writing skills, including a detailed knowledge of proper notekeeping procedures, are valuable assets when job-hunting and certainly when it comes time for that annual performance review with the boss.

How to Critique and Grade Student Notekeeping

Student notekeeping was the subject of several papers published in the *Journal of Chemical Education* between 1925 and 1930 (4–9). Precious little on the same subject has appeared since that time (10, 11). Even though some of these papers were written over 50 years ago, they addressed some of the basic concerns that teachers face today—how to assess student performance, how to motivate better perormance, and how to do these things efficiently. Today's teachers would be well advised to read these short, eloquent papers.

Most of the papers conclude that the best system for assessing student performance is to look over the notebooks during the laboratory period while the students are working. This practice serves several purposes:

1. Students are aware that they must write notes during the laboratory investigation, not later. They begin to develop a "feel" for making notekeeping habitual.

2. Students can get assistance and critical review of their work ("immediate feedback" in today's jargon.) The teacher can get a much better impression of the student's attitude and level of achievement than if the notebooks are quickly skimmed once or twice a semester.

3. The teacher does not have a thick stack of notebooks to grade at the middle and end of the semester; each student's work gets the attention that it deserves.

The single most important criterion for assessing student writing is simply, "Does the notebook clearly tell what was done?" The teacher can examine the notebook entry in comparison to an "ideal" notebook entry such as those in Chapter 6 of this book. Allowances must be made for the conditions under which the student's notekeeping is done: time constraints to finish the experiment or cooperation or the lack of it among lab partners, for example. Here, the teacher is grading the notes against an arbitrary, but achievable standard. John T. Stock, retired Professor of Analytical Chemistry at the University of Connecticut, took one of his students' finest examples of a notebook and actually chained it to a lab bench for all to see. It became the standard by which all other student notebooks were judged.

At the high school level, another method that can be both fun and frustrating is perhaps one of the best methods for showing students how important their writing really is. Students first perform experiments by following printed instructions; pairs or small groups of students each perform a different experiment. For the next class session, the notebooks are exchanged, and students then must follow directions for the experiment that were written into the notebook by their peers. This technique can be modified by an exchange of notebooks between students at different schools. This approach does raise some interesting questions, however: Can experiments be designed so that the outcome depends on how well the notes of the previous group were written? Is it fair if the student who writes good notes must read and follow the poorly written notes of another student?

Finally, grading should take into account how much the student has learned during the semester. As in some other subject areas, notekeeping skills will vary widely among students at the beginning of the semester, and smaller differences will be found among students at the end of the semester. Sophomore-level (or higher) college students can be judged by the same 10-point checklist (*see* Chapter 4) that industrial research managers use to assess employee's notebooks.

Literature Cited

1. "Undergraduate Professional Education in Chemistry: Guidelines and Evaluation Procedures;" American Chemical Society Committee on Professional Training; fall 1983.
2. Held at the University of Connecticut, Storrs, CT, 5–10 August 1984; sponsored by the ACS Division of Chemical Education, the New England Chemistry Teachers Association, and the Two-Year College Chemistry Conference.
3. Shakashiri, B. Z. *Chemical Demonstrations: A Handbook for Teachers of Chemistry*; University of Wisconsin Press: Madison, 1983.
4. Gould, H. W. *J. Chem. Ed.* 1927, *4(7)*, p. 890.
5. Mortensen, J. C. *J. Chem. Ed.* 1927, *4(7)*, p. 892.
6. Bowers, W. G. *J. Chem. Ed.* 1926, *3(4)*, p. 419.
7. Stubbs, M. F. *J. Chem. Ed.* 1926, *3(3)*, p. 296.
8. Walker, W. O. *J. Chem. Ed.* 1925, *2(6)*, p. 489.
9. Graham, H. C. *J. Chem. Ed.* 1930, *7(5)*, p. 1122.
10. Hancock, C. K. *J. Chem. Ed.* 1954, *31(8)*, p. 433.
11. Wilson, L. R. *J. Chem. Ed.* 1969, *46(7)*, p. 447.

Appendix B

Photographs from the Historical Notebooks of Famous Scientists

Late in life, Leonardo da Vinci began to write down many of the thoughts, experiments, and projects that he had undertaken throughout his extraordinary life. He wrote the note on combustion shown in Figure B.1 nearly 500 years ago, 275 years before the discovery of oxygen by Priestley. Like many of his notes, this one was written in "mirror-style" so that it appeared to be in some sort of code and could not be understood by others. If a mirror is held up to the page, the following sentence can be read.

"dove noujve lafia ma no uj ve anj mal che alitij- Il sup-chio veto vccide lafiama el tepato lanutri cha."

In Leonardo's Italian, the letters "n" and "m" were often replaced by a line over the preceding vowel, and the "u" was often written as "v" and "v" as "u". In modern Italian the quote is:

"Dove non vive la fiamma, non vive animal che aliti. Il superchio vento uccide la fiamma, e 'l temperato la nutrica."

The modern English translation is, "Where flame cannot live, no animal that draws breath can live. Excess of wind puts out flame, moderate wind nourishes it." (from Il Codice Atlantico, 270 r.a. Reproduced by permission from the Biblioteca Ambrosiana, Milan.)

The notebook pages shown in Figure B.2 record some of Leo H. Baekeland's early work on the first synthetic resin, Bakelite. He was developing a method to harden wood by impregnating strips of wood with phenol-formaldehyde condensation products. He noticed that, although the wood surface did not get hard, some material that had oozed out of the wood was very hard. He also found

0906–5/85/0129/$06.00/1

that heating some of the softer condensation products in sealed vessels could produce a hard, moldable material. In his notebook, at the bottom of page 15, Baekeland recorded his discovery that temperature plays an important role in the phenomenon of hardening. (Reproduced with permission from the Smithsonian Institution—The National Museum of American History.)

Michael Faraday's notebook entry (Figure B.3) records the discovery of electromagnetic induction, 29 August 1831. He had tried many arrangements of wires wrapped around various objects before he hit upon this observation. In his entry, he gives a detailed description of the apparatus that is supported by a clear, simple sketch. (Reproduced with permission from the Royal Institution of Great Britain.)

In Humphry Davy's notebook entry for 19 October 1807 (Figure B.4), he confirmed that potassium was a chemical element. Electrolytic decomposition of potash produced a gas, of which he wrote: "This gas proved to be pure oxygen; Capl. Expt. proving the decomposition of Potash." (Reproduced with permission from the Royal Institution of Great Britain.)

Figure B.5 contains pages from the laboratory notebook of Walter H. Brattain in which he records the discovery of the amplification effect produced by a point-contact transistor. He carefully listed the names of people who observed the experiment and included the signatures of two witnesses. (Reproduced with permission from AT&T Bell Laboratories.)

Alexander Fleming had made the serendipitous discovery of a mold that inhibited the growth of a staphylococci colony. The notebook pages shown in Figure B.6 record his first experiment in which an extract of the mold was tried against several strains of bacteria (Plate 1). He concluded, "[therefore] mould culture contains a bacteriolytic substance for staphylococci" (Plate 2). *Note* the clarity and specificity (lack of speculation) in this statement. (Reproduced with permission from the British Library and Lady Fleming.)

An early electronic computer at Harvard University (the Mark I) malfunctioned on the afternoon of 9 September 1945 (Figure B.7, Plate 1). This time, unlike previous common electrical problems, the problem was a moth caught in a relay. The notebook entry (Plate 2) made by Grace Hopper (now Commodore Hopper of the U.S. Navy) contains the insect taped to the page and reads, "First actual case of bug being found." The term *bug* was already in common use to connote problems in electrical and mechanical systems before this time. It came from the Welsh term *bwg*, meaning a bugbear or hobgoblin. (Reproduced with permission from the U.S. Naval Surface Weapons Center, Dahlgren, Va.)

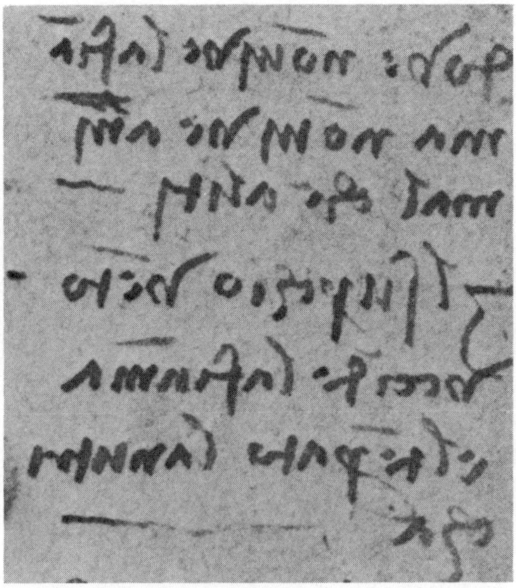

Figure B.1

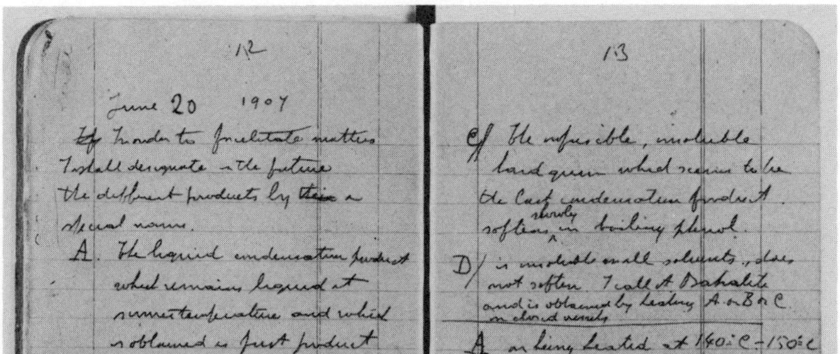

12

June 20 1907

In order to facilitate matters
I shall designate in the future
the different products by a
special name.

A. The liquid condensation product
which remains liquid at
summer temperature and which
is obtained as first product
soluble in alcohol, phenol.
acetone. Can be heated somewhat
more until it becomes pasty, but even
then is still soluble in alcohol.

B The elastic rubberlike product
which follows A. insoluble in
alcohol partially soluble in
mixture of wood alcohol & amylalcohol
soluble in acetone and in mixture
of acetone and alcohol
softens by heating

13

C The infusible, insoluble
hard gum which seems to be
the last condensation product.
softens slowly in boiling phenol.

D/ is insoluble in all solvents, does
not soften. I call it Bakelite
and is obtained by heating A or B or C
in closed vessels

A on being heated at 140°C–150°C
during 4 hours in sealed tube
becomes very hard ([illegible])
a nice smooth ivory like mass
without bubbles, amber yellow
and opaque. Has nicely taken the
inside of the tube. Insoluble in alcohol
phenol & acetone. Hardening seems
to be accompanied with slight
shrinking because the cylinder leaves
the tube very easily. Scarcely one
or two drops of watery liquid showed
inside of the tube and even this may

Figure B.2

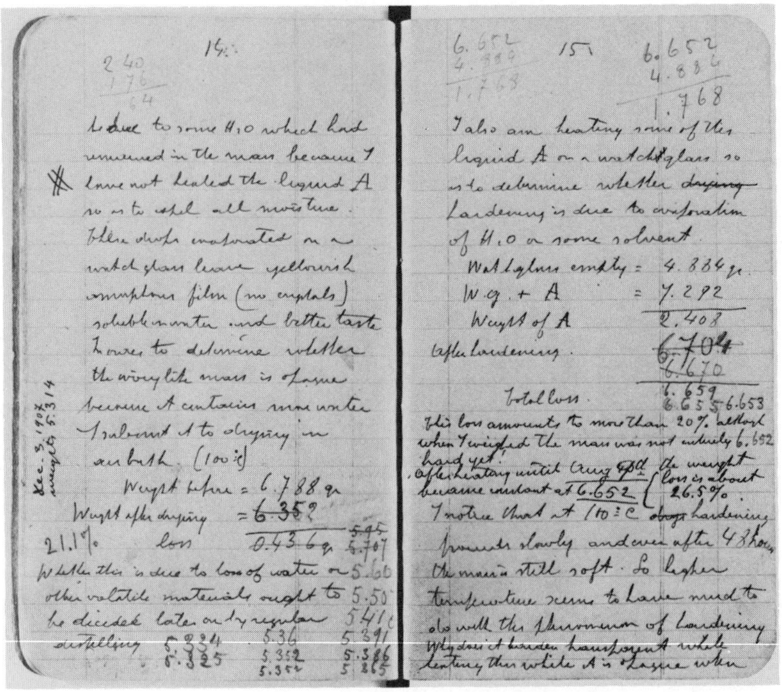

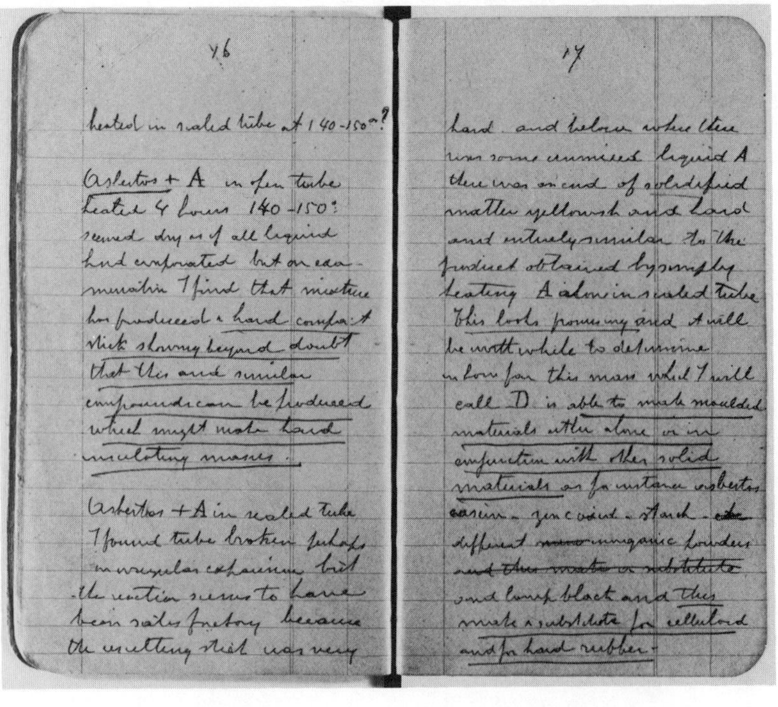

Figure B.2

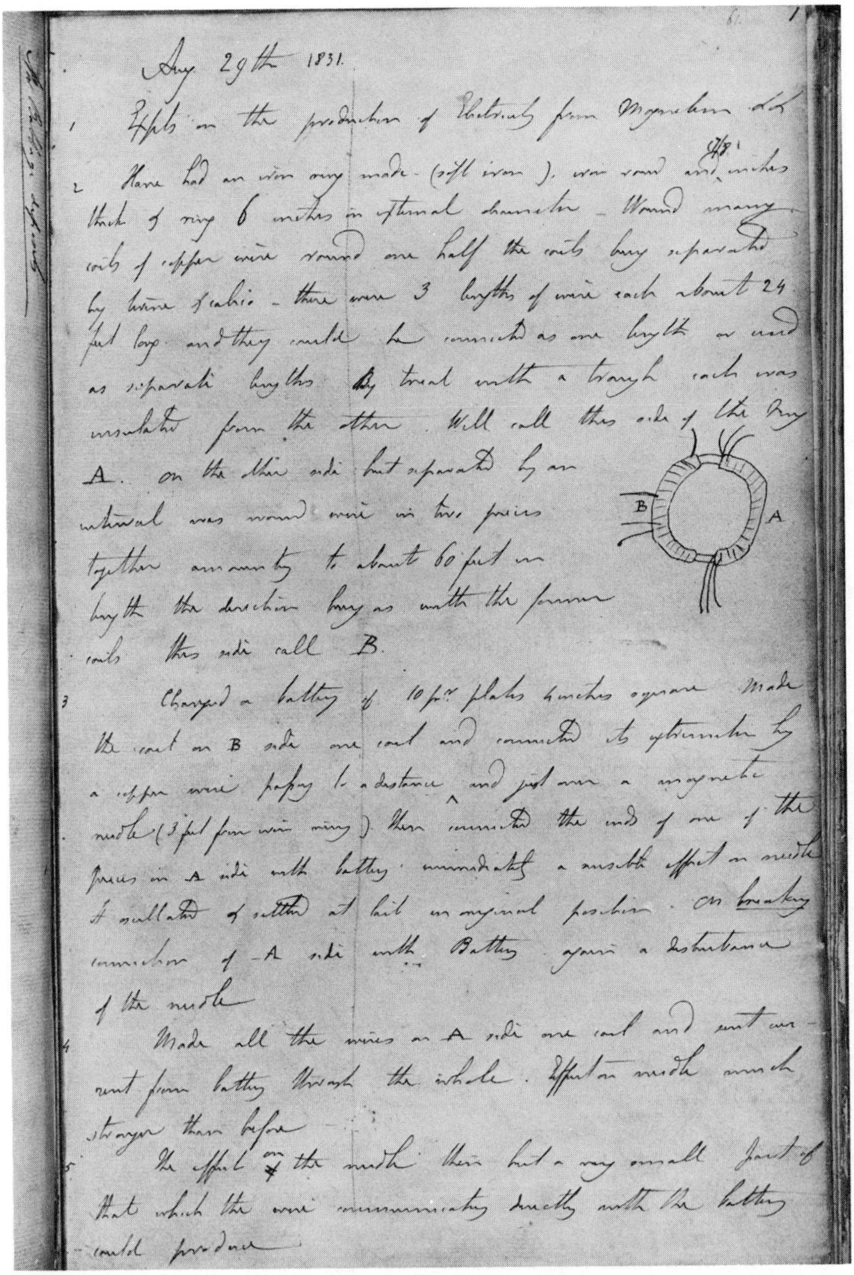

Aug. 29th 1831.

1. Expts on the production of Electricity from Magnetism, etc.

2. Have had an iron ring made (soft iron), iron round and ⁷/₈ inches thick and ring 6 inches in external diameter — Wound many coils of copper wire round one half the coils being separated by twine & calico — there were 3 lengths of wire each about 24 feet long and they could be connected as one length or used as separate lengths. By trial with a trough each was insulated from the other. Will call this side of the ring A. on the other side but separated by an interval was wound wire in two pieces together amounting to about 60 feet in length the direction being as with the former coils; this side call B.

3. Charged a battery of 10 pr plates 4 inches square. Made the coil on B side one coil and connected its extremities by a copper wire passing to a distance and just over a magnetic needle (3 feet from wire ring) then connected the ends of one of the pieces on A side with battery; immediately a sensible effect on needle. It oscillated & settled at last in original position. On breaking connection of A side with Battery again a disturbance of the needle.

4. Made all the wires on A side one coil and sent current from battery through the whole. Effect on needle much stronger than before.

5. The effect on the needle then but a very small part of that which the wire communicating directly with the battery could produce.

Figure B.3

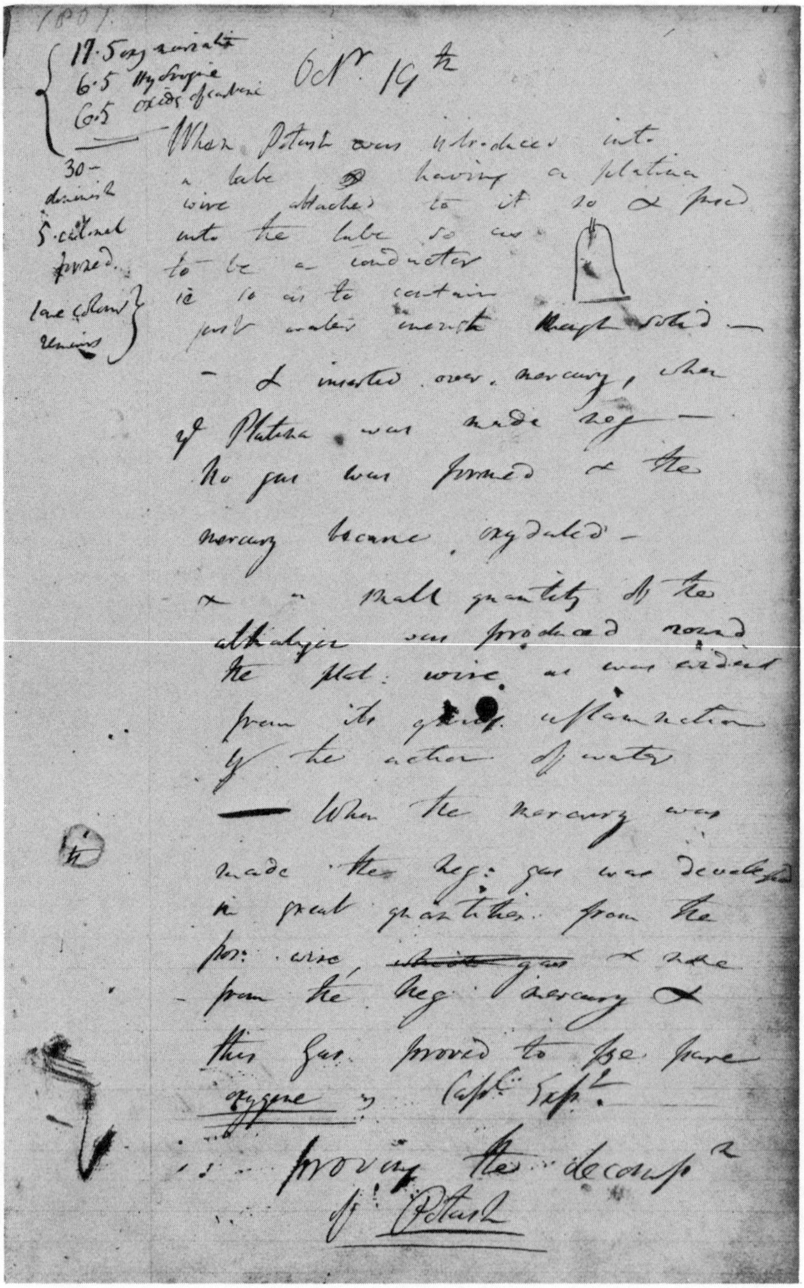
17.5 oxymuriatic
6.5 Hydrogene
6.5 oxide of carbone

Octr. 19th

When Potash was introduced into a tube having a platina wire attached to it so & fused into the tube so as to be a conductor ie so as to contain just water enough to keep it solid — & inverted over mercury, when of Platina was made neg — No gas was formed & the mercury became oxydated — & a small quantity of the alkaligen was produced round the Plat: wire as was evident from its greater inflammation by the action of water — When the mercury was made the neg: gas was developed in great quantities from the pos: wire, ~~[illegible]~~ gas & none from the neg mercury & this gas proved to be pure oxygene — Capt. Exp^t. proving the decomp^n of Potash

Figure B.4

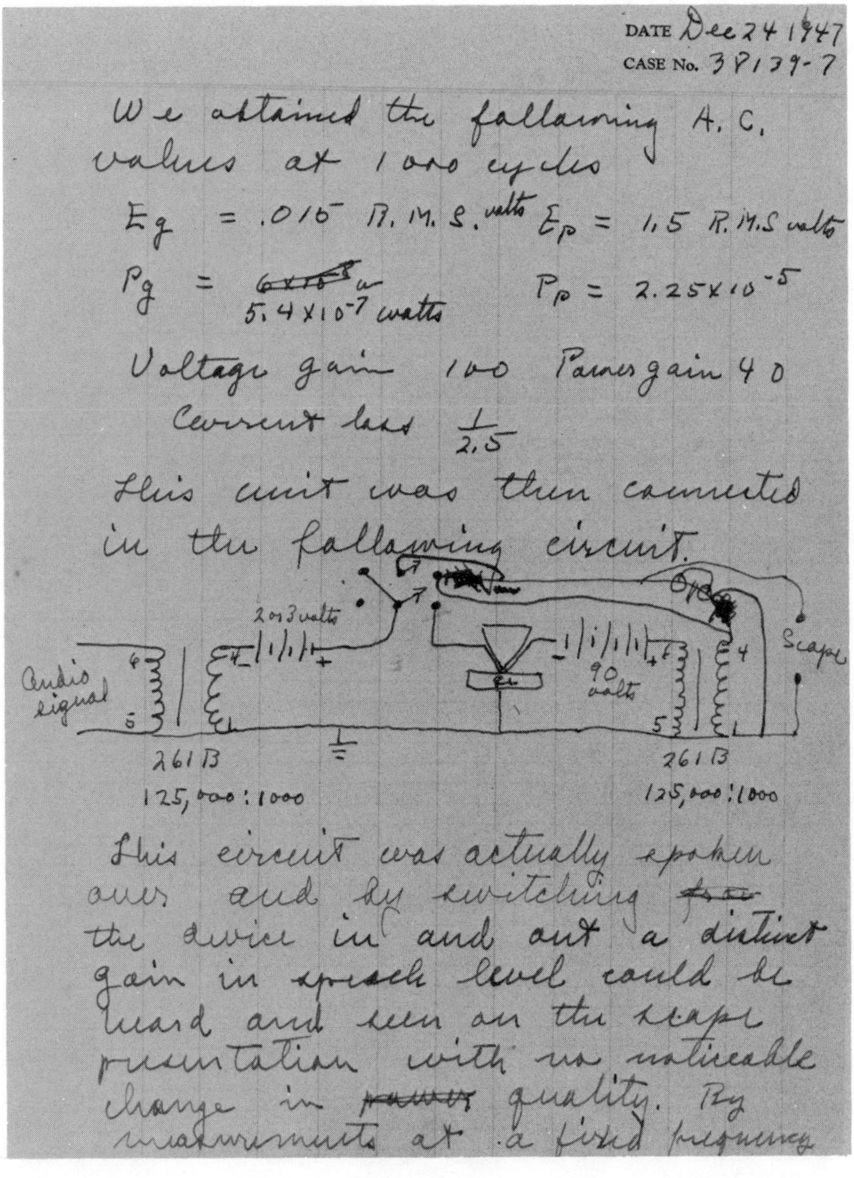

DATE Dec 24 1947
CASE No. 38139-7

We obtained the following A. C.
values at 1000 cycles

$E_g = .015$ R. M. S. volts $E_p = 1.5$ R. M. S. volts

$P_g = \dfrac{6 \times 10^{-8} w}{5.4 \times 10^{-7} \text{ watts}}$ $P_p = 2.25 \times 10^{-5}$

Voltage gain 100 Power gain 40
  Current loss $\dfrac{1}{2.5}$

This unit was then connected
in the following circuit.

This circuit was actually spoken
over and by switching ~~the~~
the device in and out a distinct
gain in speech level could be
heard and seen on the scope
presentation with no noticeable
change in ~~power~~ quality. By
measurements at a fixed frequency

Figure B.5

DATE Dec 24 1947
CASE No. 38139-7

in it was determined that the
power gain was the order of a factor
of 18 or greater. Various people
witnessed this test and listened
(were present)
of whom some were the following
R. B. Gibney, H. R. Moore, J. Bardeen
G. L. Pearson, W Shockley, H. Fletcher
R. Bown. Mrs. H. B. Moore assisted
in setting up the circuit and
the demonstration occurred on
the afternoon of Dec 23 1947

Read & understood by
G. L. Pearson Dec 24, 1947
H. R. Moore Dec 24, 1947

Dec 24 1947
This morning H. B. Moore changed
the circuit on page 7 as follows

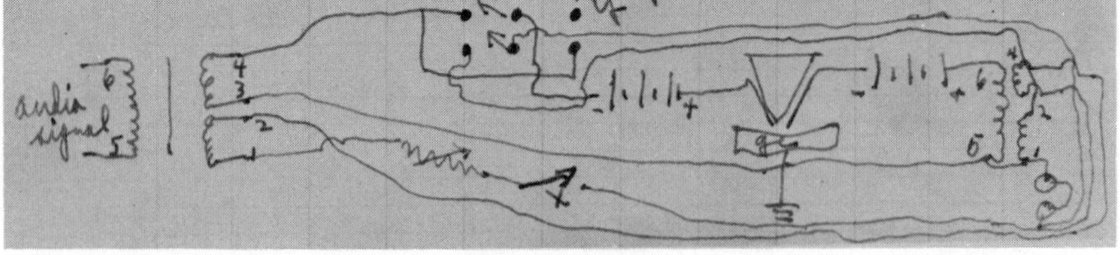

audio
signal

Figure B.5

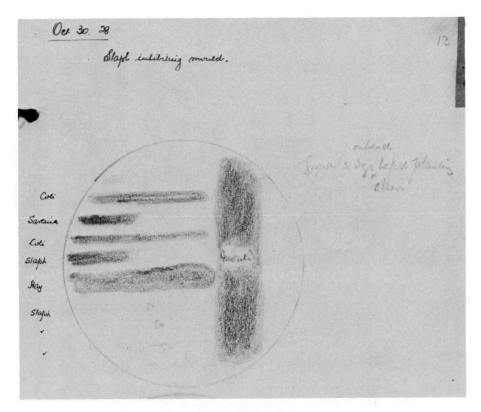

Oct 30. 28

Staph inhibiting mould.

13

Coli
Sarcina
Coli
Staph
Hay
Staph

Figure B.6, Plate 1

Oct. 30. 28.

14

Extract of staph inhibiting mould made in Ysats Centrifuged
24 hours at 37°C

To 0.5 cc of this 0.25 cc of staph emulsion added and
incubated at 45° and 56°C

In 3 hrs @ 45° considerable lysis of staph in tube with mould extract
56° no visible difference from control.

∴ mould culture contains a bacteriolytic substance for staphylococci

Figure B.6, Plate 2

Figure B.7, Plate 1

Figure B.7, Plate 2

Index

Index

Copy Editor: Susan Robinson
Cover and book design: Pamela Lewis
Production Editor: Anne Riesberg
Managing Editor: Janet S. Dodd
Typesetting: Hot Type Ltd., Washington, D.C.
Printing and binding: Maple Press Co., York, Pa.